MOLECULAR BIOLOGY INTELLIGENCE UNIT

CHROMOSOME BANDS: PATTERNS IN THE GENOME

Wendy Bickmore
MRC Human Genetics Unit
Edinburgh, Scotland, U.K.

Jeffrey Craig
Institut für Anthropologie und Humangenetik
Universität München
München, Germany

CHAPMAN & HALL
I(T)P An International Thomson Publishing Company

New York • Albany • Bonn • Boston • Cincinnati • Detroit • London • Madrid • Melbourne • Mexico City • Pacific Grove • Paris • San Francisco • Singapore • Tokyo • Toronto • Washington

LANDES BIOSCIENCE
AUSTIN, TEXAS

Molecular Biology Intelligence Unit

CHROMOSOME BANDS: PATTERNS IN THE GENOME

R.G. LANDES COMPANY
Austin, Texas, U.S.A.

Printed in the U.S.A.

Please address all inquiries to the Publishers:
R.G. Landes Company, 810 S. Church Street, Georgetown, Texas, U.S.A. 78626
Phone: 512/ 863 7762; FAX: 512/ 863 0081

North American distributor:
Chapman & Hall, 115 Fifth Avenue, New York, New York, U.S.A. 10003

CHAPMAN & HALL

U.S. and Canada ISBN: 0-412-11741-X

Library of Congress Cataloging-in-Publication Data

Bickmore, Wendy A.
Chromosome bands and the organization of the genome / Wendy A. Bickmore, Jeffrey Craig.
p. cm. — (Molecular biology intelligence unit)
Includes bibliographical references and index.
ISBN 0-57059-393-0 (alk. paper)
1. Chromosome banding I. Craig, Jeffrey. II. Title.
III. Series.
QH600.B535 1997
574.87'3282—dc20
DNLM/DLC 96-34369
for Library of Congress CIP

Publisher's Note

R.G. Landes Company publishes six book series: *Medical Intelligence Unit, Molecular Biology Intelligence Unit, Neuroscience Intelligence Unit, Tissue Engineering Intelligence Unit, Biotechnology Intelligence Unit* and *Environmental Intelligence Unit.* The authors of our books are acknowledged leaders in their fields and the topics are unique. Almost without exception, no other similar books exist on these topics.

Our goal is to publish books in important and rapidly changing areas of the biosciences for sophisticated researchers and clinicians. To achieve this goal, we have accelerated our publishing program to conform to the fast pace in which information grows in the biosciences. Most of our books are published within 90 to 120 days of receipt of the manuscript. We would like to thank our readers for their continuing interest and welcome any comments or suggestions they may have for future books.

Shyamali Ghosh
Publications Director
R.G. Landes Company

CONTENTS

FOREWORD

The ability to band chromosomes is of obvious and proven practical value in cytogenetics. In addition, the fact that chromosomes can be banded suggests that they are highly organized structures. This compartmentalization reflects fundamental aspects of chromosome function and how the nucleus handles the genome. It is this facet of chromosome banding that is pursued in this book, which draws mainly on studies of mammalian chromosomes while, where appropriate, introducing work on chromosomes from other organisms.

The subject is approached from the point of view of cell biology—an area not often considered by those who use chromosome banding in a practical way. Conversely, those interested in chromosome biology are often unaware of the lessons that can be learned from a consideration of chromosome banding patterns. Here we draw together different approaches to understanding chromosome biology, and marry lessons learned from cytology and cytogenetics—especially modern fluorescence in situ hybridization and immunofluorescence—with those from genetics and biochemistry.

This book takes a completely different standpoint from other books on chromosome banding. It concentrates not on how to band chromosomes, but rather on what the resulting patterns mean in the wider context of chromosome biology and understanding how the genome works. This is a timely consideration in view of the rapid advances in our understanding of the DNA sequence of the genome.

Acknowledgments

We would like to dedicate this book to Prof. H. John Evans to mark his retirement as Director of the Medical Research Council's Human Genetics Unit, formerly the Clinical and Population Cytogenetics Unit (CAPCU). The staff of CAPCU and Prof. Evans were at the forefront of developments in chromosome banding and chromosome identification in the 1970s. We thank them for their patience in opening the eyes of a cytogenetically ignorant molecular biologist (WB) to the beauty of chromosomes and their bands.

We thank all of our colleagues and fellow chromosome watchers for sharing our obsession. However, special thanks are due to Judy Fantes, not only for her expert cytogenetic skills but also for all things FISH related and to Paul Perry, without whom all of our microscopes and imaging systems would be worthless.

Finally we thank the Medical Research Council and the Lister Institute for Preventive Medicine for support throughout much of this work.

CHAPTER 1

The Metaphase Chromosome

Chromosomes were given their name, meaning 'colored body' in Greek, by Waldeyer in 1888,[1] a precociously apt term in light of the current techniques used in chromosome biology, so many of which employ multicolor fluorescence. These colored bodies comprise a mixture of DNA and protein that together is known as chromatin. Chromosomes are maximally condensed at metaphase of the cell cycle, when a typical mammalian chromosome will be compacted around 10,000-fold more than a DNA double helix. Hence, such chromosomes can be seen as separate entities with the light microscope. Not all scientists have been endeared by the appearance of the metaphase chromosome—this manifestation of the chromosome has even been considered to be:

> *".... the dullest form of chromosome: an inert package needed to make orderly mitosis possible."*[2]

Here we set out to refute this suggestion. We view the metaphase chromosome not only as one of biology's most visually appealing objects but also as a structure whose distinctive organization is a direct reflection of the way in which the nucleus handles the genome. It is also a structure full of pattern, and pattern in biology is usually a manifestation of underlying mechanisms of control. Thus, although the metaphase chromosome is indeed largely composed of extremely compact and inactive chromatin, it is the form in which daughter cells inherit genetic information—not only in the form of the primary DNA sequence but also as the epigenetic information that influences how the DNA sequence should be interpreted. We therefore expect to find all the components that determine chromatin memory within the structure of the metaphase chromosome.

Over 20 years ago it was discovered that certain treatments of mammalian metaphase chromosomes could reveal a series of reproducible bands across the chromosome. This had immediate practical value allowing, for example, each human chromosome to be uniquely identified. Additionally, abnormalities in the normal banding pattern that result from large scale chromosome rearrangements such as deletions, translocations and inversions could be linked with specific genetic disorders. This enabled many disease-associated genes to be mapped, and subsequently isolated by positional cloning.

The applicability of chromosome banding as a technique to investigate genome structure from a wide variety of sources of biological material (both different cell types within a species as well as different species) depends to a large extent on the ease with which metaphase chromosomes can be isolated and spread for microscopic analysis. The use of spindle microtubule poisons such as colchicine or Colcemid (demecolcine) allows cells to be accumulated in metaphase. It is a point worth bearing in mind, and one often forgotten, that although we commonly refer to chromosomes harvested in such a way as 'metaphase' they are strictly speaking, C-metaphase chromosomes (metaphase chromosomes prepared in the presence of Colcemid). This may seem pedantic, but it is unlikely that such chromosomes are the exact equivalents of chromosomes at metaphase of untreated cells.[3] In fact, they may be more analogous to chromosomes at prometaphase before congression to the metaphase plate and before tension has been applied across the kinetochore by microtubules from the spindle poles. The chemistry of proteins at the kinetochore is known to change in response to such tension and this tension cannot occur on chromosomes condensed in the presence of a microtubule poison.[4] Sister chromatids remain joined only at the centromere, not along the arms, in C-metaphase chromosomes.

Obviously, chromosome banding is not applicable as a tool in nondividing cell populations. In some tissues of some organisms, interphase chromosomes become visible after they have become polytenized, i.e., have undergone multiple rounds of replication in the absence of cell division. Polytene chromosomes can also be revealed to contain a series of bands and inter-bands but we will not consider them substantially here. Lastly, the chromosomes of simple eukaryotes are too small to be adequately resolved by light microscopy. In some species the visible metaphase chromosomes do not exhibit bands with some, or all, of the established banding techniques, and the significance of this is unclear.

Aside from practical cytogenetic considerations, the fact that metaphase chromosomes can be banded suggests that they are highly organized structures and that this compartmentalization reflects fundamental aspects of eukaryotic chromosome function. It is this facet of chromosome banding and its biological message that we wish to pursue in this book, since the techniques of chromosome banding have been comprehensively described elsewhere.[5-7] Here, we draw mostly on studies of mammalian chromosomes but where appropriate, introduce work on chromosomes from other organisms, particularly the yeasts *Saccharomyces cerevisiae* (*S. cerevisiae*) and *Schizosaccharomyces pombe* (*S. pombe*) and the dipteran insect *Drosophila melanogaster* (*D. melanogaster*).

THE HIERARCHY OF CHROMATIN PACKAGING

The 10,000-fold packing of the DNA molecule to form the chromosome at metaphase is accomplished by an ordered series of interactions with proteins (Fig. 1.1). The first and best understood of these is the formation of nucleosomes, which compacts the DNA helix 6-fold. Nucleosomes consist of 146 bp of DNA wrapped around a protein core of the histones H2A, H2B, H3 and H4 (core histones). These histones are arranged as an $(H3\text{-}H4)_2$ tetramer with two H2A-H2B dimers positioned on each face of the tetramer. The N-terminal tails of the histones that are important in determining the type of chromatin formed, protrude from the surface of the nucleosome. The 30 nm fiber, with a packing ratio of 1 in 40, arises through the folding and/or coiling of the nucleosomes. The spacing of nucleosomes along the 30 nm fiber varies between species and even between cell types within a species.[8] In one model, binding of histone H1 to the linker DNA joining

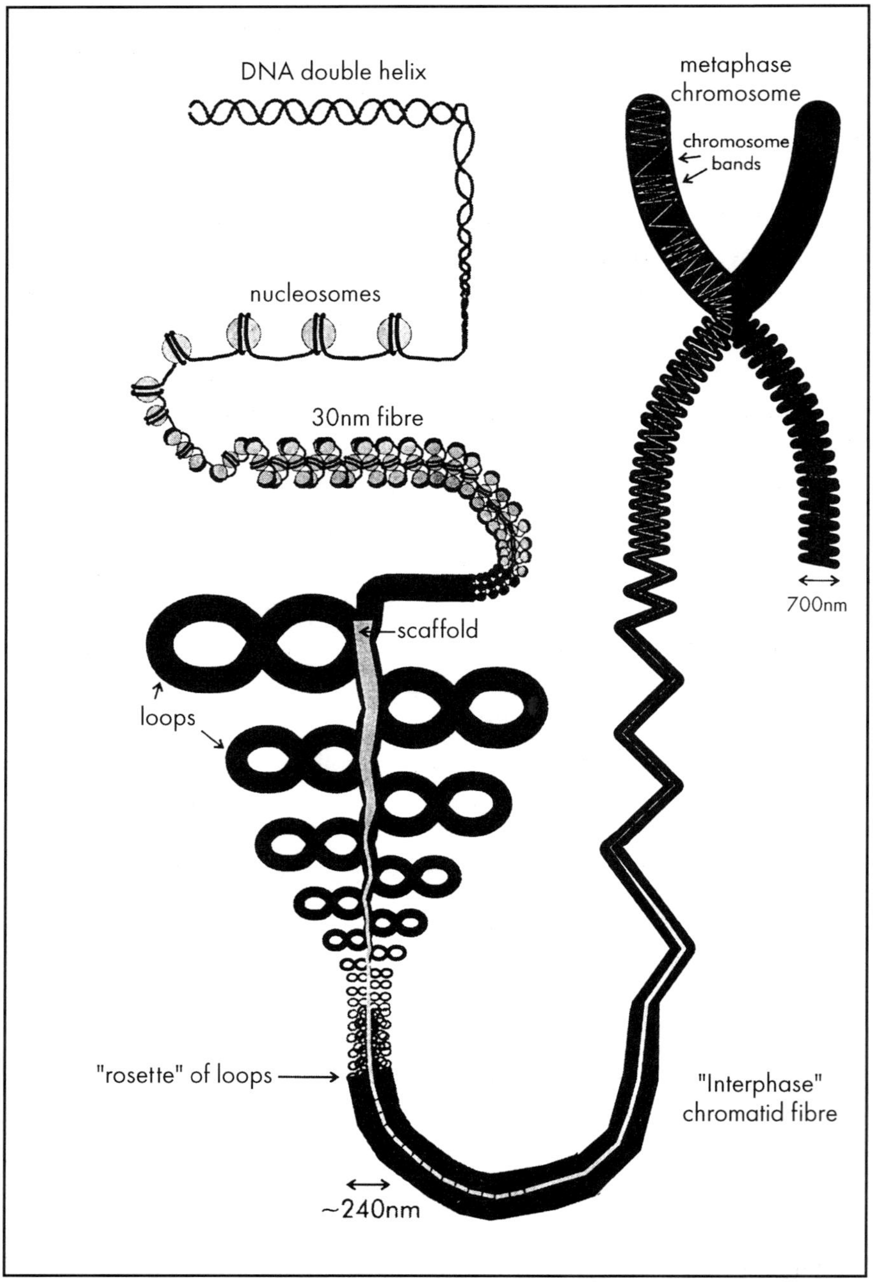

Fig. 1.1. The levels of chromatin packaging in the radial loop/scaffold model of chromosome structure. A 30 nm chromatin fiber of nucleosomes in a solenoidal arrangement[9] is shown organized into loops that radiate from a central chromosome scaffold to form a 240 nm interphase chromatid fiber.[17] This is further condensed longitudinally and laterally to form the metaphase chromatids. This is termed the radial loop/scaffold model of chromosome structure.[13-16] The path of the scaffold may be different in G- and R-bands (chapter 4).[50]

the nucleosomes arranges the nucleosomes into a regular solenoidal structure.[9] However, a study of cells at metaphase claimed to find no evidence for the presence of solenoids in the chromatin structure.[10] There are several other models in which the nucleosomes are arranged differently within the 30 nm fiber, e.g., a coiled ribbon of nucleosomes arrayed in a zigzag way.[11] Whatever the details (and these are well reviewed in ref. 12), it is accepted that something approximating to a 30 nm fiber is the basic chromatin fiber in eukaryotes.

The levels of packaging from the 30 nm fiber upwards are the least understood at the molecular and biochemical level because of the hierarchical nature of chromatin structure. However, these are the stages of packing that are most easily imagined from cytological observation of chromosomes. One model, and variations upon it, pervades and is summarized in Figure 1.1.[13-16] In this radial loop/scaffold model, the 30 nm fiber is formed into loops that are attached at their bases to a nonhistone protein "scaffold." The diameter of this fiber is approximately 250 nm and it may be the final level of packing of the interphase chromosome—hence it has been termed the chromatid fiber.[15]

The final level of chromosome packaging occurs as interphase chromosomes enter mitosis or meiosis. The longitudinal shortening of the chromosome has been suggested to occur through a helical coiling of the central chromosome axis along the chromosome's length.[15] Lateral contraction might be envisaged as a condensation of the radial loops in toward the center of the chromosome through folding, twisting or coiling. At metaphase, the chromosome scaffold can be revealed and visualized by microscopy after extraction of most proteins from the chromosomes.[18] This reverses the lateral compaction of the radial loops, allowing them to be seen[25] spread out and emanating from the central scaffold. Extraction may also destroy the longitudinal coiling of the scaffold (Fig. 1.2). The most abundant protein constituents of the mitotic chromosome scaffold are DNA topoisomerase II (topo II) and ScII. The role of these proteins in organization and condensation of the chromosome is further discussed in chapter 4.

Another strong visual piece of evidence in favor of a radial loop/scaffold-type model is the appearance of the meiotic lampbrush chromosomes of newt cells.[17] However, this vision of chromosome structure is by no means the only one. There are models that are variations on the radial loop/scaffold. For example, groups of loops may make a complete turn about the scaffold in one plane to form a disc-like 'mini-band'—a very structured and regular way to put

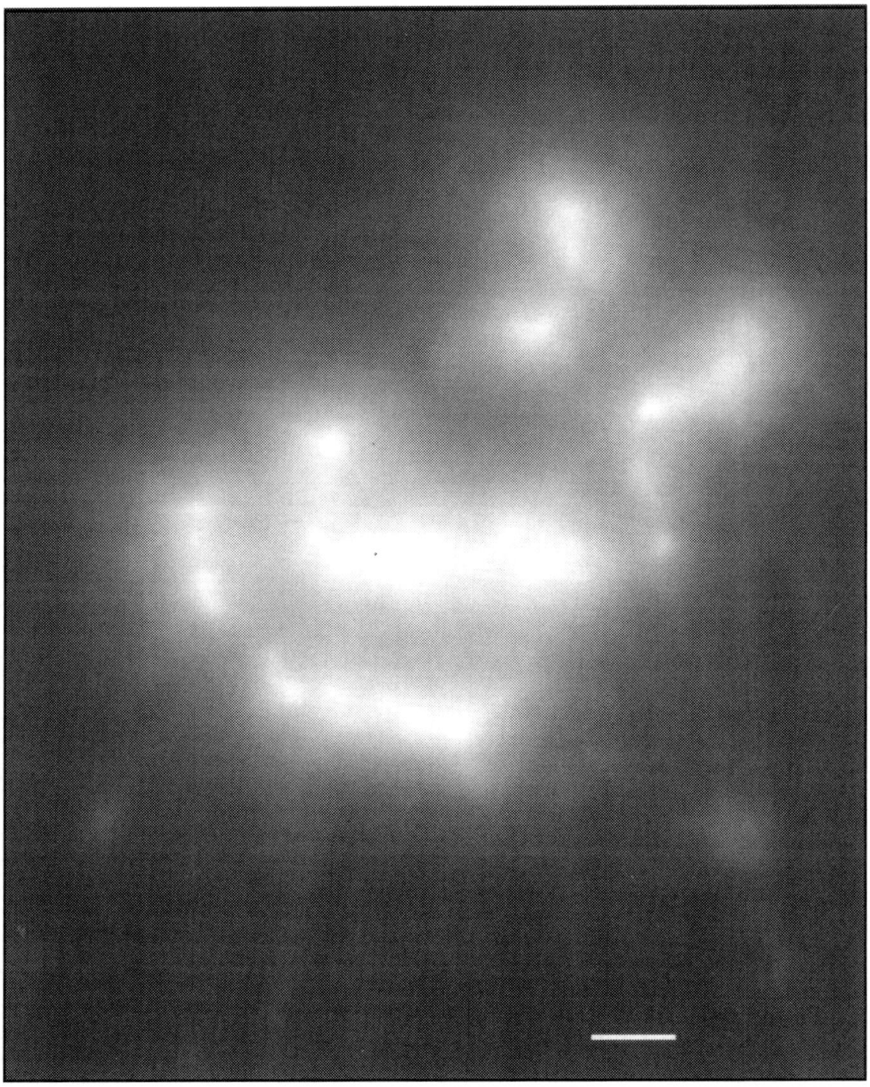

Fig. 1.2. Decondensed metaphase chromosomes. Two human C-metaphase chromosomes decondensed with 1M NaCl and stained with DAPI. Note the symmetry of the cores of the sister chromatids. DNA extends out from the chromosome cores for 3-4 μm. Bar = 2 μm.

the chromosome together.[20,21] Models that deviate most from the concept of the radial loop/scaffold are those that envisage a folded chromatin fiber which has no central chromosome axis and no anchored loops of chromatin.[22,23] The image that these models tend to suggest is of a more random and less specific and controlled chromosome structure. At the extremes, rather than the DNA molecule essentially starting at one end of the chromosome and being folded up consecutively until the other end of the chromosome is reached, the DNA and chromatin fiber could be doubled back on themselves along the chromosome length so that the base sequence and chromosome length need not be colinear.

Three-dimensional reconstruction of unfixed and unstained human metaphase chromosomes by electron microscopy indicates that chromosomes do not have a hollow core. This has been suggested to argue against the existence of a helically coiled and rigid central structure. Nor was there any evidence for a particularly dense core in the chromosome, except at the position of the centromere.[24] All of these pieces of evidence are probably most consistent with a model of chromosome structure in which there is a distributed or dispersed and noncontiguous chromosome scaffold or axis upon which 30 nm fiber loops are constrained, with a folding and coiling of both the loops and the axis superimposed on this. It is possible that the chromosome axis need not be located rigidly at the center of the chromosomes nor that the organization of the loops be radially symmetric. It is probably important, in the absence of definitive experimental evidence, not to get too caught up in the details of one model or the other at this stage.

THE FATE OF THE METAPHASE CHROMOSOME IN THE NUCLEUS

The purpose of metaphase is to segregate the genetic material accurately to daughter cells or gametes. The fate of metaphase chromosomes is therefore to become decondensed and fulfill their functions as interphase chromosomes. How different is the interphase state from that at metaphase? The extent to which chromosomes condense on passing from interphase to metaphase and decondense again in the next interphase differs among eukaryotes. In the yeast *S. pombe* there may be only a 2-fold compaction of the chromatin,[25] while a ratio of 9:1 has been calculated for compaction of mammalian metaphase chromosomes over their interphase counterparts.[14]

Textbooks generally portray an image of the interphase chromosome as a loose ball of spaghetti spread across the nucleus. However, fluorescence in situ hybridization (FISH) of probes for individual whole chromosomes to human interphase nuclei, by a process known as chromosomal in situ suppression hybridization (CISS), paints distinct globular domains (Fig. 1.3) with a large degree of higher order packaging remaining.[26-28] Therefore individual DNA sequences within a single chromosome are not free to adopt random positions within the nucleus, they are constrained to some extent by the behavior of the whole chromosome. For example, a sequence located close to rDNA on the human acrocentric chromosomes (#13, 14, 15, 21, 22) cannot help but find itself in the physical proximity of the nucleolus.[29] Therefore, the organization of sequences along the metaphase chromosome influences the location of that sequence at interphase, and all of the possible functional consequences that that may entail. Maximal spatial separation between any two points on a single human chromosome at interphase has been estimated to be ~4.5 μm. However, on a finer scale DNA sequences that are within a few Mb of each other behave as if they are part of a completely flexible chain and the spatial separation between them can be described by a simple statistical relationship—the random walk model.[30]

Do whole chromosomes adopt random positions within the nucleus? Apart from the exception of meiosis, homologous chromosome pairs of mammals do not generally seem to associate with each other either within the nucleus, or on the metaphase plate. There can be quite extreme spatial segregation of chromosomes derived from different parental sources at both interphase and metaphase,[31-33] however, there may also be cell-cycle regulated pairing of loci subject to genomic imprinting.[34] This is in contrast to the situation in *Drosophila* where homologous chromosomes are usually paired at interphase and can be shown to interact genetically.[35] *Drosophila* embryo prometaphase chromosomes adopt a 'Rabl-like' configuration with the telomeres clustered at one side of the nucleus and centromeres located at the opposite pole. Again,

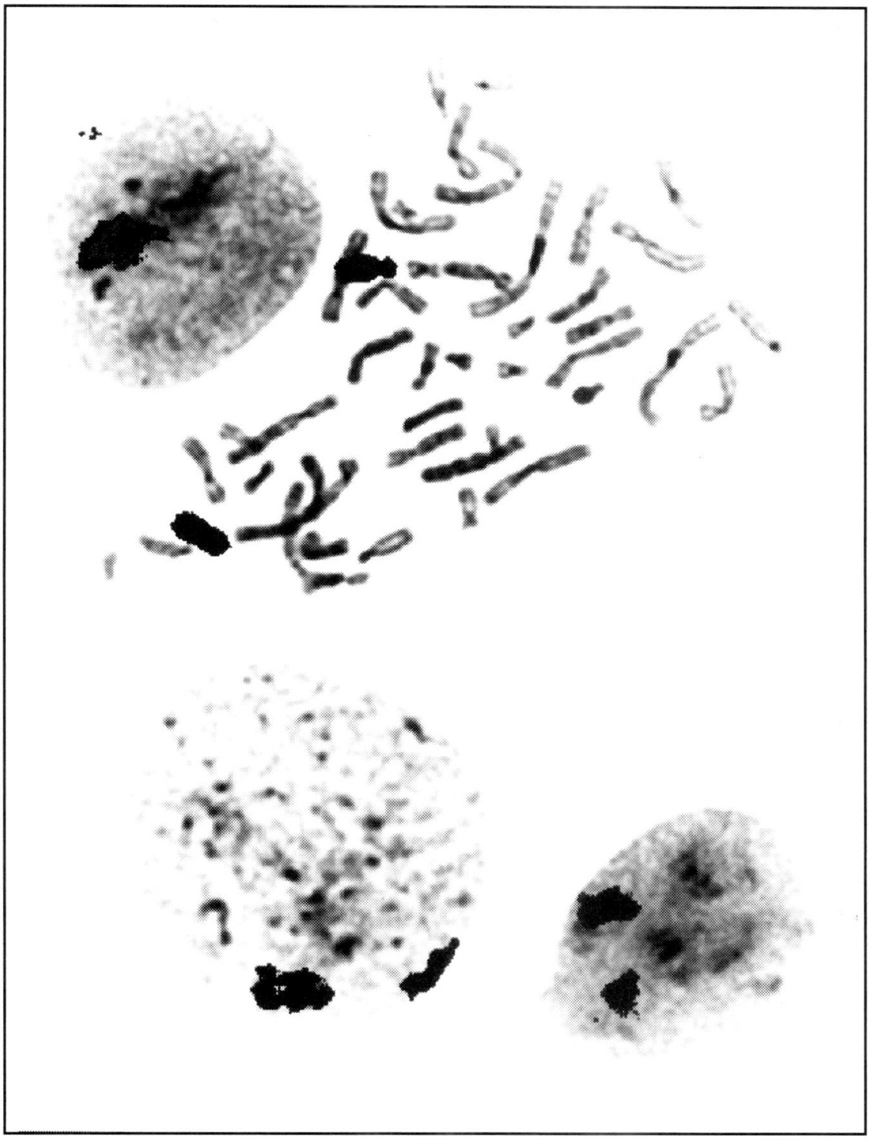

Fig. 1.3. Interphase and metaphase chromosomes. Three human interphase nuclei and a metaphase spread (top right) from stimulated peripheral lymphocytes stained with DAPI (G-positive bands are dark) and hybridized to a probe for chromosome 18 (black). The two chromosome 18s painted black can be seen in the metaphase spread as can the very compact chromosome domains in the interphase nuclei.

there is no consistent evidence that this is the way mammalian chromosomes are typically organized,[36] but it has been suggested that specific chromosomes are spatially positioned relative to each other in a systematic way.[33] The positions of mammalian centromeres and telomeres relative to the nuclear periphery may vary during the cell cycle.[32]

FUNCTIONS OF THE METAPHASE CHROMOSOME

Why is a condensed chromosome needed for cell division? One simple answer might be that it is needed to endow the 240 nm interphase chromatid fiber with sufficient strength to maintain its integrity during the mechanics of cell division at which time the chromosomes may be subject to a variety of physical forces pushing and pulling on the chromatids.[37] Condensation also serves as an aid to the separation of sister chromatids, driving the equilibrium of topoisomerase II activity toward decatenation of the duplicated DNA molecules.[38] A condensed chromosome may also be needed to provide the correct chromatin template on which to assemble a fully functional kinetochore that attaches to mitotic spindle microtubules and maintains sister chromatid attachments until anaphase.

The condensed chromosome has been considered as having a number of distinctive domains or surfaces.[17] The first is the central domain, where the chromatin fiber is packed close to the chromosome axis, core or scaffold. Several specialized proteins are associated with the chromosome scaffold, and specific sites along the chromosomal DNA sequence find themselves in this domain and may influence the activity of the chromosome in the subsequent interphase.[39] The central domain of the chromosome at the site of the kinetochore is characterized by several proteins needed to form the functional kinetochore.

A second domain is on the outside of each chromatid where the sister chromatids contact each other. This has been called the pairing domain and is present both along the arms and at the centromeres, however, the nature of the association in these two regions differs. Colcemid arrested chromosomes have detached arms but attached centromeres, and in untreated chromosomes the arms of sister chromatids can be seen to separate at late prophase whereas splitting at the centromere is not seen until anaphase.[40] In order for successful and accurate chromosome segregation sister chromatid attachments must be maintained during metaphase and be strong enough to resist the forces transmitted through spindle pole microtubules and yet be swiftly released at the metaphase/anaphase transition. Some of this cohesion can simply be envisaged as coming from catenated sister DNA molecules and being released by a population of late acting topoisomerase II molecules located perhaps at the centromeres.[41] However, in addition to this it is clear that some of the controlled release of sister chromatids operates through proteins that help to hold sister chromatids together and through ubiquitin-mediated proteolysis of these proteins at the metaphase/anaphase transition.[42] Specific proteins have been localized to the pairing domain between sister chromatids of mammalian chromosomes. CLiPs (chromatid linking proteins) are present throughout the length of the pairing domain, whereas INCENPs (inner centromere proteins) are confined to the centromere where they are among the most tightly bound chromosomal proteins known until anaphase when they detach from the chromosomes and move to the metaphase plate.[43,44] However, the precise functions of these two types of protein is unclear—though it has been suggested that the INCENPs have no real chromosomal function and are just hitching a ride on the chromosomes. Recently, proteins required for correct sister chromatid cohesion and separation in mitosis and meiosis have been identified.[45,46] Unlike mitotic metaphase, during metaphase I of meiosis, sister chromatid attachments are maintained at centromeres and not released until anaphase of the second division.

Lastly, the metaphase chromosome has a surface domain, of which the kinetochore region is a specialized form—as may be the telomeres at the ends of the chromosome. The surface domain is the region of the chromosome that is most exposed to other cellular components. Since specific sites within the DNA associate with the central domain of the chromosome, it follows that specific sites must also find themselves on the chromosome surface. What proteins are found at the chromosome surface? A specialized set of proteins are found at distinct sites within the trilaminar structure of the kinetochore including

INCENPs A-F.[47] A variety of other proteins have been detected over the surface of metaphase chromosomes but excluded from the kinetochore region, including components of the nucleolus, the nuclear envelope, and ribosomes as well as splicing components.[48]

Aside from the kinetochore that fulfills a very obvious and vital chromosomal function, what might be the role, if any, of the surface of the mitotic chromosome? As has been suggested in the case of the INCENPs, the chromosome may merely act as a passive carrier of some proteins to specific sites in the daughter cells. Since the nuclear membrane breaks down during mitosis in most eukaryotes this could be a way of ensuring that proteins find their way into the new nucleus of daughter cells rather than being distributed randomly to both nucleus and cytoplasm. This is most likely to the case for components of the nucleolus such as fibrillarin, B23 and Ki67, that are found on the surface of chromosomes. These proteins may also serve to protect the chromosomes from exposure to cytoplasmic components upon nuclear envelope breakdown as cells enter mitosis.

Mitotic chromosomes are essential for the formation of the new nucleus that begins late in anaphase or in telophase. The chromosome surface acts as a template upon which the components of the newly forming nuclear envelope assemble. Integral membrane proteins and proteins that associate with lamins associate first, followed by the lamins themselves, and these interactions are modulated by mitotic phosphorylation.

Lastly, there is an emerging model of chromosome structure at interphase in which the surface of chromosome domains form contacts with a nuclear matrix that plays a central role in transcription and RNA processing.[27] While the relationship between the exterior surface of a chromosome at metaphase and interphase is not clear, it is likely that they are related topologically in some way.

Does the whole chromosome mass or the inside or outside of chromosomes contribute to banding patterns? Common themes of folding and coiling recur within any discussion of chromosome structure—where does chromosome banding come into these themes of coils within coils? Different techniques of chromosome banding may reveal the differing organization of blocks of the genome at the level of the primary DNA sequence, nucleosome structure or aspects of the folding into loops radiating from a chromosome scaffold. These different and multifaceted levels of chromosome structure are dealt with in subsequent chapters.

References

1. Waldeyer W. Über Karyokinese und ihre Bezeihungen zu den Befruchtungsvorgängen. Arch. Mikrosk. Anat. 1888; 32:1. English translation by Benham WB in Q. J. Microscop. Sci. 1889; NS 30:159-278.
2. Crick F. Postscript to the sixth international chromosome conference in Helsinki, Finland. In: de la Chapelle A and Sorsa M, eds. Chromosomes Today. Vol. 6. Amsterdam: Elsevier North Holland Biochemical Press, 1977:403-408.
3. Rieder CL and Palazzo RE. Colcemid and the mitotic cycle. J. Cell Sci. 1992; 102:387-392.
4. Gorbsky GJ. Kinetochores, microtubules and the metaphase checkpoint. Trends Cell Biol. 1995; 5:143-148.
5. Verma RS and Babu A. Human chromosomes: Principles and Techniques. McGraw Hill, 1995.
6. Sumner AT. Chromosome banding. London:Unwin Hyman, 1990.
7. Gosden JR. Chromosome Analysis Protocols. In: Walker JM, ed. Methods in Molecular Biology. Vol 29. Series Humana Press, 1994.
8. Alegre C and Subirana JA. The diameter of chromatin fibers depends on linker length. Chromosoma 1989; 98:77-80.
9. Finch JT and Klug A. Solenoidal model for superstructure in chromatin. Proc. Natl. Acad. Sci. USA 1976; 73:1897-1901.
10. McDowell AW, Smith JM and Dubochet J. Cryo-electron microscopy of vitrified chromosomes in situ. EMBO J. 1986; 5:1395-1400.
11. Woodcock CL, Grigoryer SA, Horowitz RA and Whitaker N. A folding model for chromatin that incorporates linker DNA variability produces fibers that mimic the native structures. Proc. Natl. Acad. Sci. USA 1993; 90:9021-9025.
12. Wolffe A. Chromatin: Structure and Function. Academic Press, 1995.
13. Marsden MPF and Laemmli UK. Metaphase chromosome structure: evidence for a radial loop model. Cell 1979; 17:849-858.
14. Rattner JB and Lin CC. Radial loops and helical coils coexist in metaphase chromosomes. Cell 1985; 42:291-296.
15. Boy de la Tour E and Laemmli UK. The metaphase scaffold is helically folded: sister chromatids have predominantly opposite helical handedness. Cell 1988; 55:937-944.
16. Earnshaw WC. Mitotic chromosome structure. BioEssays 1988; 9:147-153.
17. Rattner JB. Integrating chromosome structure with function.

Chromosoma 1992; 101:259-264.

18. Paulson JR and Laemmli UK. The structure of histone-depleted metaphase chromosomes. Cell 1977; 12:817-828.
19. Callan HG. Lampbrush chromosomes. Proc. R. Soc. Lond. (B) 1982; 214:417-448.
20. Pienta KJ and Coffey DS. A structural analysis of the role of the nuclear matrix and DNA loops in the organization of the nucleus and chromosome. J. Cell Sci. 1984; 1(S):123-135.
21. Manuelidis L and Chen TL. A unified model of eukaryotic chromosomes Cytomet. 1990; 11:8-25.
22. Mullinger AM and Johnson RT. Packing DNA into chromosomes J. Cell Sci. 1980; 46:61-86.
23. Belmont AS, Braunfeld MB, Sedat JW and Agard DA. Large-scale chromatin structural domains within mitotic and interphase chromosomes in vivo and in vitro. Chromosoma 1989; 98:129-143.
24. Harauz G, Borland L, Bahr GF et al. Three-dimensional reconstruction of a human metaphase chromosome from electron micrographs. Chromosoma 1987; 95:366-374.
25. Umesono K, Hiraoka Y, Toda T and Yanagida M. Visualisation of chromosomes in mitotically arrested cells of the fission yeast *Schizosaccharomyces pombe.* Curr. Genet. 1983; 7:123-128.
26. Manuelidis L and Borden J. Reproducible compartmentalization of individual chromosome domains in human CNC cells revealed by in situ hybridisation and three-dimensional reconstruction. Chromosoma 1988; 96:397-410.
27. Cremer T, Kurz A, Zirbel R et al. Role of chromosome terrritories in the functional compartmentalization of the cell nucleus. Cold Spring Harbor Symp. Quant. Biol. 1993; 52:777-792.
28. Trask BJ, Allen S, Massa H et al. Studies of metaphase and interphase chromosomes using fluorescence in situ hybridisation. Cold Spring Harbor Symp. Quant. Biol. 1993; 53:767-775.
29. Kaplan FS, Murray J, Sylvester JE et al. The topological organization of repetitive DNA in the human nucleolus. Genomics 1993; 15:123-132.
30. Yokota H, van den Engh G, Hearst JE et al. Evidence for the organization of chromatin in megabase pair-sized loops arranged along a random walk path in the human G0/G1 interphase nucleus. J. Cell Biol. 1995; 130:1239-1249.
31. Leitch AR, Schwarzacher T, Mosgoller W et al. Parental genomes are separated throughout the cell cycle in a plant hybrid. Chromosoma 1991; 101:206-213
32. Ferguson M and Ward DC. Cell cycle dependent chromosomal movement in pre-mitotic human T-lymphocyte nuclei. Chromosoma 1992; 101:557-565.

33. Nagele R, Freeman T, McMorrow L and Lee H-Y. Precise spatial positioning of chromosomes during prometaphase: evidence for chromosomal order. Science 1995; 270:1831-1835.
34. LaSalle JM and Lalande M. Homologous association of oppositely imprinted chromosomal domains. Science 1996; 272:725-728.
35. Tartof KD and Henikoff S. Trans-sensing effects from *Drosophila* to humans. Cell 1991; 65:201-203.
36. Hiraoka Y, Agard D and Sedat JW. Temporal and spatial coordination of chromosome movement, spindle formation, and nuclear envelope breakdown during prometaphase in *Drosophila melanogaster* embryos J. Cell Biol. 1990; 111:2815-2828.
37. Fuller MT. Riding the polar winds: chromosomes motor down east. Cell 1995; 81:5-8.
38. Holm C. Coming undone: how to untangle a chromosome. Cell 1994; 77:955-957.
39. Bickmore WA and Oghene K. Visualizing the spatial relationship between defined DNA sequences and the axial region of metaphase chromosomes. Cell 1996; 84:95-104.
40. Sumner AT. Scanning electron microscopy of mammalian chromosomes from prophase to telophase. Chromosoma 1991; 100:410-418.
41. Shamu CE and Murray AW. Sister chromatid separation in frog egg extracts requires DNA topisomerase II activity during anaphase. J. Cell Biol. 1992; 117:921-934.
42. Holloway SL. Sister chromatid separation in vivo and in vitro. Curr. Opinion Genet. Develop. 1995; 5:243-248.
43. Rattner JB, Kingwell BG and Fritzler MJ. Detection of distinct structural domains within the primary constriction using autoantibodies. Chromosoma 1988; 96:360-367.
44. Mackay AM, Eckley DM, Chue C and Earnshaw WC. Molecular analysis of the INCENP's (inner centromere proteins): separate domains are required for association with microtubules during interphase and with the central spindle during anaphase. J. Cell Biol. 1993; 123:373-385.
45. Stratmann R and Lehner CF. Separation of sister chromatids in mitosis requires the Drosophila pimples product, a protein degraded after the metaphase/anaphase transition. Cell 1996; 84:25-35.
46. Kerrebrock AW, Moore DP, Wu JS and Orr-Weaver TL. Mei-S332: A Drosophila protein required for sister chromatid cohesion can localize to meiotic centromere regions. Cell 1995; 83:247-256.
47. Pluta AF, Mackay AM, Ainsztein AM et al. The centromere: hub of chromosomal activities. Science 1995; 270:1591-1601.
48. Hernandez-Verdun D and Gautier T. The chromosome periphery during mitosis. BioEssays 1994; 16:179-185.

49. Foisner R and Gerace L. Integral membrane proteins of the nuclear envelope interact with lamins and chromosomes, and binding is modulated by mitotic phosphorylation. Cell 1993; 73:1267-1279.
50. Saitoh Y and Laemmli UK. From the chromosomal loops and the scaffold to the classic bands of metaphase chromosomes. Cold Spring Harbor Symp. Quant. Biol. 1993; 153:755-765.

CHAPTER 2

Chromosome Banding Techniques

Below we describe briefly some of the ways in which chromosome bands can be produced. This is intended to provide a taste of the types of procedure involved rather than detailed protocols. There are several excellent treatises that describe in detail the methodology of chromosome banding (for example refs. 1-3). Of the panoply of banding techniques we will concentrate principally on those methods that differentiate the euchromatic arms of mammalian chromosomes. We describe not only the techniques used in practical cytogenetics (G-, R-, Q- and replication banding) but also a variety of more esoteric methods which have been used mainly as tools to try and unravel aspects of the structure of the metaphase chromosome. In this chapter we mention only briefly techniques that highlight specifically either centromeres, heterochromatin or nucleolar organizers, though the molecular properties of heterochromatin are discussed extensively in chapter 6.

WHAT IS A CHROMOSOME BAND?

A number of specific stains identify metaphase chromosomes as uniform structures. However, some staining methods produce species-specific transverse patterns of differentially-stained bands. These banding patterns are, in the main, invariant during development and between cells from different tissues of the same species. Banding protocols can be dye-based or function-based. Hence a chromosome band is a manifestation of a segment of chromatin that has relatively homogeneous and distinctive functional and structural features over a large enough stretch for it to be visually distinguished from its neighboring segments. Techniques used to reveal chromosome bands are considered here enhance an inherent pattern of chromosome organization that occurs in domains large enough to be seen by light microscopy.

Dye-based chromosome banding techniques range from the simple (i.e., no pretreatments and one type of dye molecule), to the more complicated, which require pretreatment of the chromosomes prior to staining and require combinations of dyes. The most common methods of dye-based chromosome banding are C- (centromere), Q- (quinacrine), G- (Giemsa) and R- (reverse) banding. These methods can all be used to band mammalian chromosomes, but each individual method has its own limitations among vertebrate species. Bands that show a stronger staining reaction should strictly be referred to as positive bands; conversely, weakly- or nonstaining bands are negative bands. However, G-positive bands will usually be termed G-bands and likewise for R-positive bands (R-bands). Positive C-bands contain constitutive heterochromatin. For practical purposes euchromatic Q-positive bands can be considered to be identical to G-positive and R-negative bands.

The most widely used function-based chromosome banding protocol is replication banding. It is based on the facts that eukaryotic DNA replication is under temporal control and that different bands replicate their DNA at different times during S phase of the cell cycle. As a general rule the DNA in R-bands is replicated earlier than G-bands. The chromomere patterns observed on meiotic chromosomes might also be considered as a function based form of chromosome banding. G-bands correspond to the condensed meiotic chromomeres and R-bands to the interchromomeric regions. Comparisons of human chromosome 9s banded with a variety of techniques are shown in Figure 2.1.

C-BANDING

C-banding involves sequential acid and alkali treatment of chromosomes, incubation in hot standard saline citrate (SSC), followed by staining with Giemsa.[4,5] Giemsa is a dye mix consisting of red eosin Y and blue thiazines such as azure B and, without pretreat-

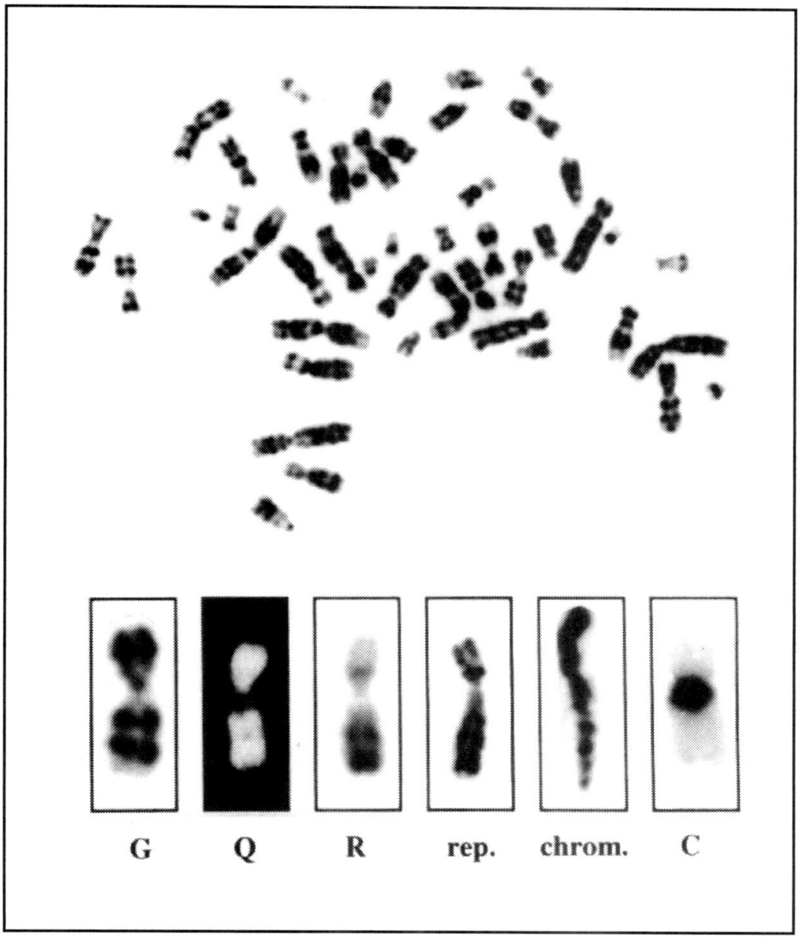

Fig. 2.1. Examples of human chromosome 9 banded by: G, trypsin G-banding (G-positive bands dark); Q, Q-banding (Q-positive bands bright); R, R-banding with heat and Giemsa (R-positive bands dark); rep., replication banding, with early replicating (R-band) regions staining dark; chrom, pachytene chromomeres; C, C-banding. Adapted from ref. 95.

ment, stains mammalian metaphase chromosomes uniformly purple. Positive C-bands often identify constitutive heterochromatin, which contains highly repetitive satellite DNA. Constitutive heterochromatin in humans includes centromeres (α satellite) and other noncentromeric regions, including satellites I, II and III on chromosomes 1, 9, 16 and Y.[6,7] These regions contain no known active genes in mammals and banding is thought to result from the selective retention of DNA in C-bands and loss of non C-band

DNA during banding pretreatments. In *Drosophila,* however, some genes do reside in heterochromatin and their location there is important for their correct function—this is discussed further in chapter 6.

The inert nature of the chromatin in C-bands suggests that there may be an interaction between DNA and proteins in heterochromatin that is distinct from, or more extreme than, that occurring in euchromatic regions. Indeed, C-bands appear to be protected against digestion with nucleases such as DNase I by the presence of tightly bound nonhistone proteins that restrict access of the enzyme to the DNA.[8] In vivo these same proteins may have a role in restricting the access of endogenous enzymes such as RNA polymerase to heterochromatic parts of the genomes. The specialized chromosomal proteins that can be found in heterochromatin are discussed in chapter 6.

G-BANDING

G-banding involves staining with Giemsa dye mixture and has evolved as a technique from two different sources. The first of these was a modified form of C-banding in 2 x SSC at 60°C, in which G-bands are sometimes stained weakly;[5] indeed heterochromatin in plants is sometimes confusingly referred to as G-band DNA. G-banding patterns are usually identical to those produced by the fluorescent dye quinacrine, except for differences in heterochromatin staining. A complete G-banded karyotype of the human genome is shown in Figure 2.2. Pretreatment of chromosomes with agents such as proteases, urea, SDS, β-mercaptoethanol and permanganate that might be expected to digest, denature, reduce or oxidize proteins, respectively, also produced G-bands. Initial protocols used either pronase and Giemsa[9] or trypsin and Leishman's stain (like Giemsa and Wright's a Romanowski stain).[10] All Romanowski dyes are mixtures of eosin Y with thiazine dyes. G-bands have also been produced using direct staining in solutions of Giemsa in alkaline buffer.[11]

G-banding is thought to result from interactions of both DNA and protein with the thiazine and eosin components of the stain, but the detailed chemistry is poorly understood. There are more arguments for than against the conclusion that G-banding

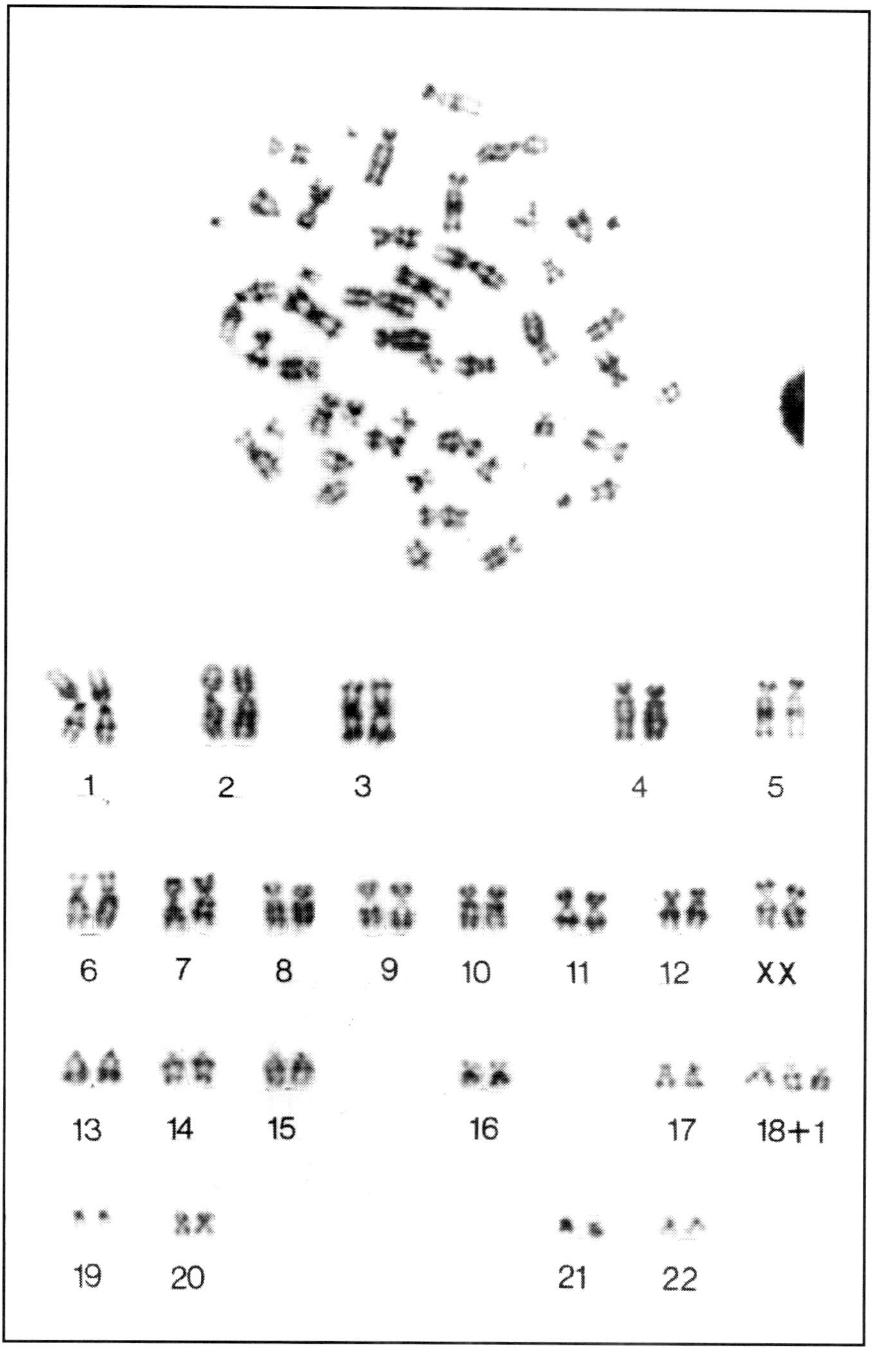

Fig. 2.2. A G-banded spread and karyotype analysis of human metaphase chromosomes. G-banded (G-bands are dark) peripheral lymphocyte chromosomes from a female with trisomy 18.

treatments do not induce major morphological changes in chromosome structure, and that they do not induce differential availability of dye binding sites on chromosomal DNA.[2] An alternative hypothesis has been suggested[12] in which biochemical interactions between charged Giemsa dye molecules and DNA, the core histone octamer and histone H1 of fixed chromosomes are invoked. The need for some reorganization of chromatin during G-banding is supported by the absence of banding in chromosomes that have been fixed with aldehyde cross-linking agents.[2] A comparison of light microscope (LM) and scanning electron microscope (SEM) pictures of trypsin-Giemsa banded human chromosomes found that lateral chromosome "grooves" identified with the SEM corresponded to the regions identified as G-bands with the LM. The conclusion reached was that Giemsa molecules were more likely to be trapped in the groove created by the digestion of G-bands with trypsin, but no explanation could be given as to why trypsin preferentially digested these bands.[13,14]

Despite the fact that G-banding is poorly understood mechanistically, it is one of the most widely used banding procedures. In a purely practical sense, not understanding the basis for a staining procedure does not hamper its use in the laboratory but it is dissatisfying from the point of view of trying to understand the basic chromosome organization that is being revealed by the technique.

R-BANDING

R-, or reverse (to G) banding is also a commonly employed banding procedure. The most common method involves partial heat denaturation of chromosomes in hot (87°C) acidic saline followed by Giemsa staining.[15] This method is thought to result in the preferential denaturation of AT-rich DNA, followed by the staining of the under-denatured GC-rich regions, but as G-banding, the role of the various dye components in the resulting staining pattern is not clear.[2]

In some R-banding protocols Giemsa dye mixture has been replaced with acridine orange, giving green/yellow fluorescent R-bands and orange/red G-bands.[16] Acridine orange is thought to have two modes of DNA binding; the first by intercalation giving green fluorescence which can only occur in regions of double-stranded DNA, whereas external binding to the phosphate backbone results in red fluorescence and can take place on single-stranded DNA. Hence acridine orange mediated R-banding may reflect the preferential denaturation of G-band DNA due to differences in base composition between R- and G-bands (chapter 3). A role for denaturation of DNA is supported by the observation that at high temperatures acridine orange stains chromosomes wholly red, while at low temperatures the staining is green. Additionally, it has been reported that in situ hybridization of DNA probes to chromosomes (a procedure that requires a denatured DNA template) can be performed on chromosomes that have been R-banded but not otherwise denatured.[17] Acridine orange also bands chromosomes that have been subject to photolysis in the presence of dyes, such as methylene blue, that may crosslink protein and DNA or destroy G residues in the DNA sequence.

The complementarity of R- and G-banding patterns (positive G-bands are negative R-bands and vice versa) suggests that these two banding techniques may be revealing reciprocal manifestations of the same underlying basic phenomenon in the chromosome.

T-BANDING

T-banding identifies a subset of R-bands—the most intensely staining ones—by employing either a more severe heat treatment than used in R-banding, or by a combination of dyes or fluorochromes.[18,19] It is thought to identify the GC-richest R-bands, of which approximately half occur at telomeres in the human genome,[20] hence the name. The mechanism involved is thought to be similar to that of R-banding: denaturation of all but the most GC-rich DNA. SEM studies revealed that T-bands consisted of fibrous, 3-dimensional regions of chromatin and all interbands appeared flat; an indication that material appeared to have been lost from these latter regions during the T-banding procedure.[21]

Because T-bands are often telomeric it has been postulated that all T-bands may have origins at the ends of chromosomes.[19] There is little evolutionary evidence to support this idea; regions of the mouse genome in conserved synteny linkage groups to human T-bands mostly occupy nonterminal parts of mouse chromosomes (chapter 3). It remains to be seen whether these regions exhibit banding properties similar to the analogous human T-bands.

Q-BANDING

Q-banding was the first method described for banding the euchromatic arms of human and plant chromosomes, in 1968.[22] The technique involved staining with the fluorescent dyes quinacrine or quinacrine mustard, which react specifically with certain bases, and was based on the premise that the base composition along chromosomes must be nonhomogeneous. This led to the full karyotypic analysis of the human chromosome complement, allowing each of the 23 pairs of chromosomes to be differentiated and identified.[23] Q-positive bands are also G-positive bands (some heterochromatic regions are also stained—one of the most well known being the block at the end of the long arm of the human Y chromosome), but Q-banding is more widespread in evolution than G- or R-banding and so G- and Q-bands are not strictly equivalents.

Quinacrine molecules intercalate into chromosomal DNA irrespective of sequence, but fluoresce brighter in regions of AT-rich DNA. By combining Q-banding with tritiated precursor labeling it was shown that the brighter a chromosome segment stained with quinacrine the greater its AT content.[24] Interspersion of guanine quenches quinacrine fluorescence and may also affect the excitation efficiency.[25] Hence it is generally accepted that Q-positive bands are more AT-rich than Q-negative bands. Dimeric analogs of quinacrine such as spermine bis acridine $(CMA)_2S$ have practical advantages over quinacrine itself since they produce crisper, more long lasting banding patterns.

BANDING WITH OTHER FLUOROCHROMES

The distinction between R-banding with acridine orange and other forms of fluorochrome banding is that the former depends on pretreatment of the chromosomes, whereas the latter does not. There are a number of other molecules whose fluorescence is influenced by the base composition of the DNA to which they are bound. In addition to quinacrine, other fluorochromes with a specificity for AT-rich DNA include; Hoechst 33258 (2,2-[4-hydroxyphenyl]-6-benzimidazolyl-6-[1-methyl-4-piperazyl]-benzimidazol-trihydrochloride), DAPI (4'-6-diamidino-2-phenylindole), DIPI (4'-6-bis[2'-imidazolinyl-4'-5'-H]-2-phenylindole), dibutyl-proflavine, daunomycin and adriamycin. For most of these dyes their fluorescence is not quenched by guanine and thus they give less distinct bands than does quinacrine.[2] However, daunomycin fluorescence is greatly quenched by DNA with a GC content of >32%. Unlike quinacrine, daunomycin and adriamycin, Hoechst and DAPI are not thought to intercalate but to bind in the minor groove of the helix.[26] DAPI staining has the advantage that it is very resistant to fading.

Fluorochromes with a preference for GC-rich DNA include the chromomycin antibiotics: chromomycin A_3, mithramycin and olivomycin, and 7-amino actinomycin D—a fluorescent derivative of actinomycin D. These dyes give an R-band like pattern.

Dyes can also be used in combination to band chromosomes, exploiting the competition for binding sites on the chromosomal DNA by the dyes. In this case only one of the two dyes need be fluorescent. The counterstain may have the same or opposing base preferences to the primary stain. Commonly used combinations include distamycin A enhanced DAPI banding, actinomycin D and DAPI, DAPI and propidium iodide (PI). Nonfluorescent actinomycin D competes with DAPI for DNA binding sites in R-bands whereas nonfluorescent distamycin A competes for AT-rich DNA.[27] DAPI staining has also been combined with chromomycin A3 and mithramycin to give complementary banding patterns that are dependent on the wavelength of the exciting light. The use of stains and counterstains in banding is covered comprehensively in reference 1.

Another way in which dyes can be used in combination to give enhanced banding is through fluorescence energy transfer (FRET). When a "donor" fluorescent dye and an "acceptor" dye (which need not be fluorescent) are in close proximity, the acceptor absorbs light emitted by the donor, modifying its banding pattern. A good example of an acceptor is methyl green, a nonfluorescent molecule with specificity for AT-rich DNA. In combination with chromomycin A_3, actinomycin D or YOYO-1, it produces an R-banding pattern.[27,28]

It is generally accepted that fluorochrome banding reflects some aspects of the variation in base composition of DNA along chromosome length (discussed further in the next chapter). However, we should not lose sight of the fact that proteins in the local chromatin environment might also influence both the accessibility of the dyes to the DNA and the excitation of and emission from fluorescent dyes. Indeed, recently some new and intriguing ideas about how some fluorochrome banding patterns might reflect intrinsic aspects of chromosome structure have been raised (chapter 4).[28]

REPLICATION BANDING

Different regions of the eukaryotic genome replicate at different times during S phase, reflecting the links that must lie between the organization of the genome and the temporal control of the DNA replication machinery. In fact it has even been suggested that G- and R-bands correspond to chromosomal clusters of synchronously replicating replicons.[29] The relationship between timing of replication and chromosome banding has been studied by incorporating pulses of thymidine analogs, either tritiated thymidine[30] or 5-bromo-2'-deoxyuridine (BrdU)[31,32], into living cells during defined stages of S phase, and examining chromosomes in the subsequent metaphase. Sites of BrdU incorporation can be detected indirectly since they have decreased staining with Hoechst 33258, Giemsa, acridine orange or DAPI. This may be due to UV-induced DNA-protein crosslinks, photolysis, or altered interactions with chromosomal proteins in regions with BrdU-substituted DNA. More directly, the sites of BrdU incorporation are detected with antibodies that detect the presence of BrdU in denatured DNA.[33]

The period of S during which BrdU is incorporated into replicating DNA is controlled most precisely through the use of agents that interfere with the normal progression of DNA synthesis (methothrexate, excess thymidine or BrdU itself, fluorouracil or fluorodeoxyuridine [FdU]). These blocks can then be released in the presence or absence of BrdU. To get an early replication pattern, high concentrations of BrdU are used as the blocking agent and then cells are released from the block in the absence of BrdU.

For late replication patterns BrdU is introduced after the block is released. High levels of thymidine or BrdU act through feedback inhibition of the enzyme ribonucleotide reductase, and because of the complex allosteric controls exerted on this enzyme by nucleotides, the end result is decreased levels of free dCTP and an inhibition of DNA synthesis (addition of deoxycytidine bypasses this block as it is rapidly phosphorylated to dCTP).[34] Methotrexate and FdU reduce levels of thymidine; the first through competitive inhibition of dihydrofolate reductase—the enzyme that produces tetrahydrofolate, and the second through inhibition of thymidylate synthetase—the enzyme that uses tetrahydrofolate as the source of the methyl group to convert dUMP to dTMP. In fact, all of these agents are thought to arrest cells at two stages of the cycle—G1/S and also midway through S. Since metaphase chromosomes are harvested only four to five hours after release from the replication block only the banding pattern on chromosomes from cells arrested at the latter stage are examined (cells blocked at G1/S will not have had enough time to progress to metaphase).[33] The block to replication in the middle of S phase has been called the 3C pause[29] and it is not clear whether this is a cell cycle checkpoint in normal cells whose replication has not been perturbed or whether it is an artifact induced by the inhibitors.

All of these studies of chromosomal replication have shown that each chromosome band begins and completes its replication in a discrete short period of S. Approximately 50% of the genome, that is contained in R-bands, replicates early in S phase whereas G-band DNA replicates late (Fig. 2.3).

At a few sites in the genome, lateral asymmetry in BrdU incorporation between sister chromatids is seen. These sites occur at the location of simple satellite sequences that have an A/T strand bias (BrdU is a dT analog) such as in the heterochromatin of the long (q) arm of the human Y chromosome.[35]

CHROMOSOME BANDING AND FLUORESCENCE IN SITU HYBRIDIZATION

Currently, one of the major demands for chromosome banding is to identify the sites of hybridization of probes after fluores-

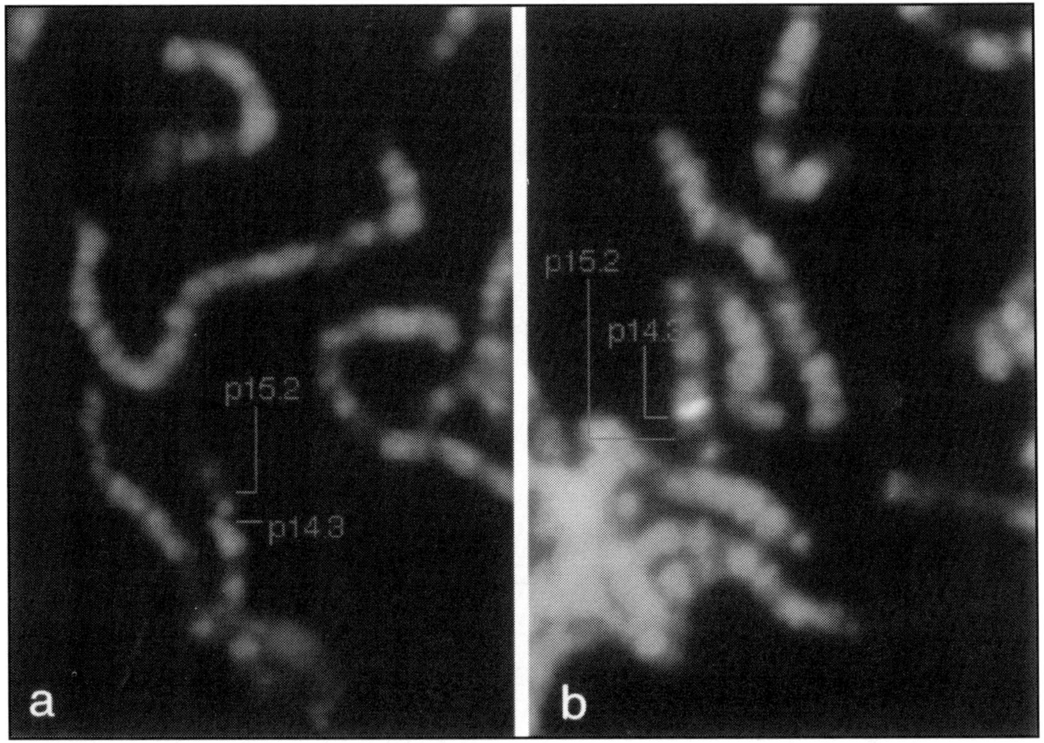

Fig. 2.3. FISH and high-resolution banding. FISH signal (red or orange) from YACs mapping to human chromosome 11 is detected on chromosomes stained with DAPI (blue) and in which BrdU has been incorporated into late replicating DNA (green). (a) The YAC maps to 11p15.1 (an early replicating R-band) and hence the hybridization signal is red since there is no BrdU present in this region of the chromosome. (b) The YAC maps to 11p14.3 (a late replicating G-band) and so the hybridization signal appears orange where red is superimposed on green (from the underlying BrdU). Reproduced courtesy of Judy Fantes, MRC Human Genetics Unit, Edinburgh. See color figure in insert.

cence in situ hybridization (FISH). Therefore, some form of fluorescence-based banding is the method of choice, although pre- or post in situ Giemsa based banding procedures have also been used but require careful registration of the banded chromosomes with the hybridization signal.[36]

A variety of fluorochrome banding techniques can be used with FISH, as can replication banding, with BrdU incorporation being detected with a fluorescently conjugated anti-BrdU antibody after hybridization (Fig. 2.3).[37] The use of fluorochromes for banding in combination with FISH is dictated not only by the quality of the bands but also by the excitation and emission spectra of the fluorochromes used. They must be compatible with the fluorochromes used in probe detection (often fluorescein or rhodamine derivatives) and with the filters fitted to microscopes.[1] Commonly used DNA counterstains in FISH include DAPI, PI and Hoechst 33258—either alone or in combination with each other. The banding on images collected with digital cameras can also be enhanced electronically and this can aid considerably in karyotype analysis (for example, compare panels A and B of Figure 2.4). Other successfully reported combinations of dyes that give good banding after FISH have included chromomycin A3/distamycin A,[38] Hoechst 33258/PI,[39] BrdU with PI or Hoechst 33258.[40,41] Chromosome banding using DNA probes is discussed in a later section.

MEIOTIC CHROMOMERES

Chromosomes at the pachytene stage of meiosis can be stained, without any pretreatment, to give a pattern of densely staining, condensed "chromomeres" separated by less condensed inter-chromomeric DNA. Each meiotic chromosome has a characteristic chromomere pattern that corresponds to the metaphase banding patterns, with chromomeres being equivalent to mitotic G-bands—they stain brightly with DAPI but not with chromomycin[42,43] with the exception of some telomeric chromomeres that are R-bands at mitotic metaphase.[19] Other R-band regions of the genome correspond to interchromomeric regions of meiotic chromosomes and hence are elongated with respect to the condensed G-band regions on such chromosomes as compared to their mitotic counterparts.[44]

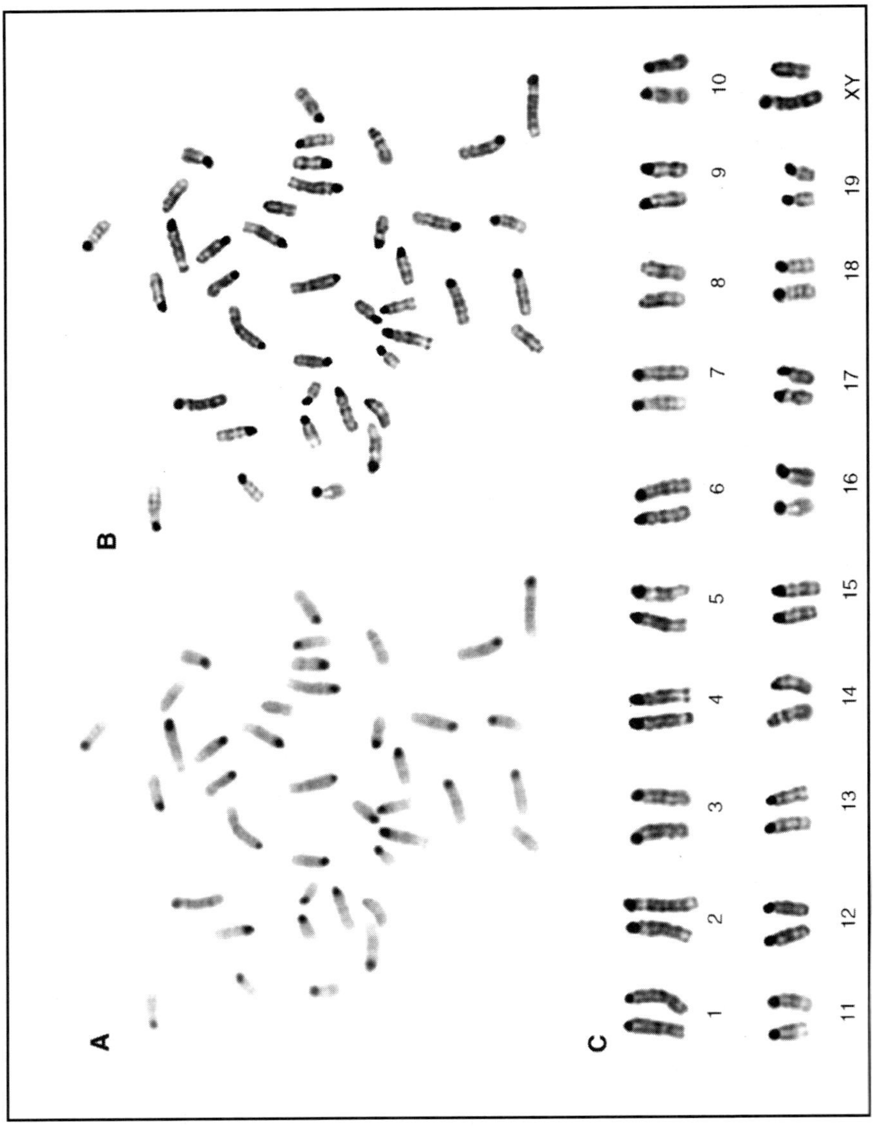

Fig. 2.4. Enhanced DAPI banding. Metaphase chromosomes from a male mouse embryonic stem (ES) cell line derived from the 129 strain of mouse stained with DAPI after FISH. The image was collected with a CCD camera and the original color image split into its component parts. The DAPI staining (originally colored blue) was converted to a monochrome image so that G-bands appear dark (A). (B) The same image is shown after electronic enhancement of the banding pattern. (C) Karyotype analysis of B. Karyotype analysis courtesy of Muriel Lee, MRC Human Genetics Unit, Edinburgh.

BANDING WITH NUCLEASES

The chromatin of active genes is generally considered to be more accessible to nuclease attack than is inactive chromatin. Aspects of these differences appear to be maintained on fixed mitotic chromosomes. DNase I sensitive sites on mitotic chromosomes, visualized by nick translation with labeled nucleotides from the cut sites, appear to correspond to R-bands in the main[45] and especially the telomeric T-bands in the human karyotype.[46] However, DNase I digestion of chromosomes is hard to control, being very sensitive to the enzymatic activity. Overdigestion of mitotic chromosomes removes all DNA except that at the pericentromeric heterochromatin. Other nucleases also preferentially digest R-bands and T-bands of intact chromosomes, with G-bands and C-bands being increasingly refractive to digestion.[2,47,48] There is controversy as to how results from restriction enzyme digestion of chromosomes should be interpreted. As with DNase I treatment, the resulting banding patterns vary according to the extent of digestion and studies using identical enzymes have given different results. Restriction enzymes with GC-rich recognition sequences always show increased digestion in R bands, which may reflect their higher GC content or greater accessibility, and enzymes with CG dinucleotides in their recognition sites often highlight T-bands.[49] Studies using enzymes with AT-rich recognition sites have shown preferential digestion in G-bands, R-bands, or no particular band type at all.[49,50] Enzymes with frequent and neutral recognition sequences often leave only C-bands intact—reflecting their nuclease resistant chromatin structure (chapter 6). Worst of all, isoschizomers can produce differing results on the same chromosomes.[51] The best that we can conclude from these experiments is that the banding patterns produced by restriction enzymes are certainly not merely related to the distribution of the enzyme's recognition sequence on the chromosome.

BANDING WITH ANTIBODIES

A variety of different chromosome banding patterns can be generated by reaction of metaphase chromosomes with antibodies against either the DNA or protein components of the chromosome. These approaches are almost entirely research tools for the investigation of chromosome structure rather than routine banding procedures. The protein epitopes detected by several of the antibodies used are destroyed or removed by standard methanol: acetic acid fixation and so unfixed hypotonically swollen cell suspensions are usually spread by cytocentrifugation onto slides, then cell membranes permeabilized with nonionic detergents to allow antibodies access to the chromosomes.[52] Slides are fixed, usually in formaldehyde, after incubation with both primary and secondary antibodies, and can be subsequently processed for FISH to allow colocalization of protein and DNA on mitotic chromosomes.[53]

Denatured chromosomes react with anti-nucleoside specific antibodies to produce chromosome banding patterns that are consistent with fluorochrome banding patterns and the differences in base composition of different bands (reviewed in ref. 2). Anti-A and anti-T bind preferentially to G-bands, and anti-C binds preferentially to R-bands. Furthermore, an antibody to double-stranded stretches of poly(dGdC) identifies R-bands.[54] An antibody directed against the modified base 5-methylcytosine (5-MeC) highlights constitutively heterochromatic regions of mammalian chromosomes[55], suggesting that these are the sites in the genome where this base is most concentrated.

There are a number of other antibodies that specifically highlight areas of heterochromatin on mammalian chromosomes. These include antibodies directed against the methylated CpG-binding protein MeCP2[56]—as expected from the distribution of 5-MeC, and against the chromo domain containing protein M31[57], a member of a protein family known to be involved in heterochromatin formation. These patterns reflect the differences in the extent of DNA modification and in the type of chromosomal proteins bound in heterochromatin compared to euchromatin. There are several antibodies that specifically detect either centromeres or telomeres—these are discussed in chapter 6.

G-band-like staining patterns have been noted with antibodies against the chromosomal protein HMG-I/Y,[58] and with an antibody that is said to recognize histone H1.[59] In contrast, antibodies directed against the most highly acetylated forms of histone H4 produce an R-band pattern on human chromosomes.[60]

Antibodies have been raised against unusual forms of DNA. G- and C-bands of mammalian chromosomes appear to be highlighted by antibodies directed against triplex DNA though we do not know whether this form of DNA actually occurs in native chromosomes.[61] Fluorochromes such as Hoechst, DAPI, daunomycin and adriamycin may only be able to bind the B-form of double-stranded DNA helix. The presence of other conformations of DNA in chromosomes could therefore conceivably contribute to banding patterns generated with these dyes.[62,63] In vitro, double-stranded DNA can also form a left-handed helix—Z-DNA—under conditions of elevated salt concentration.[64] Antibodies raised against Z-DNA highlight regions of GC-rich heterochromatin on mammalian metaphase chromosomes and methylation of cytosines may stabilize Z-DNA.[65] As with triplex DNA, this does not mean that Z-DNA is present in normal chromosomes—extraction of proteins during fixation may drive some B-form DNA to Z-DNA.[66,67] It is not known whether the other forms of right-hand DNA helices, A- and C-forms, exist in vivo.

BANDING BY DNA HYBRIDIZATION

Banding patterns that arise as a consequence of differences in the nature of the primary DNA sequence along chromosomes can be generated by FISH. The first example of this was the R- and G-banding patterns produced on human metaphase chromosomes by the Alu and LI interspersed repeats, respectively.[68] A similar nonrandom distribution of interspersed repeat families has been found in the mouse genome, suggesting that during mammalian evolution repeats of the SINE family have had a preference to integrate into R-band DNA, whereas LINE repeats seem to prefer G-band DNA.[69] Both SINE and LINE probes have been used to generate banding patterns to aid the fine localization of single-copy probes to mammalian chromosomes by FISH.[70,71]

Fractions of DNA of differing base composition (isochores) also produce banding patterns on metaphase chromosomes with the most GC-rich fractions highlighting T-bands.[72] Finally, a fraction of DNA (CpG islands) that contains the 5' ends of approximately 50% of mammalian genes also produces a T-banding pattern on chromosomes which indicates that genes are not uniformly distributed along the chromosome's length.[73] Variations in the characteristics of the primary DNA sequence along chromosomes that correspond with banding patterns is the subject of the next chapter.

Chromosome banding by FISH with DNA probes is set to take a leap forward in the near future. Chromosome-specific 'paints' can be derived from DNA prepared from flow sorted chromosomes or from somatic cell hybrids.[74] Chromosome paints are just FISH hybridization probes that consist of a complex mixture of DNA sequences that will hybridize to and highlight large visible domains of chromosomes as opposed to the single chromosomal sites detected by single locus FISH probes. The entire human karyotype can now be discriminated by multicolor FISH using chromosome paints labeled with combinations of six fluorochromes and specialized narrow band pass filters that allow discrimination between them[75] or by detailed spectral analysis with an interferometer of the fluorescence emitted through wide band pass filters by combinations of a few fluorochromes—so called spectral karyotyping (SKY).[76] Chromosome-specific R-band paints can be generated by inter-Alu PCR of DNA from somatic cell hybrids, or by purification of CpG islands from individual chromosomes.[70,73] Now that much of the human genome is cloned and mapped in yeast artificial chromosomes (YACs), pools of YACs can be selected that will cover a specific chromosome region, or set of bands. By combining these sets, labeled in multiple colors, custom-made chromosome banding patterns can be generated for any chromosome[77] and even for the entire chromosome set.[78] Since these type of 'banding patterns' are entirely determined by the experimental design and are independent of any underlying compartmentalization of the genome we consider it best that they be distinguished from classical chromosome bands—'chromosome bar codes'[77] is a more appropriate term for this next generation of ways of segmenting the genome visually.

G11-BANDING

G11-banding is performed by Giemsa staining at elevated pH (pH 11). It is a noteworthy approach because it is able to distinguish between primate and rodent chromosomes within the same cell, i.e., in somatic cell hybrids.[79] It is an enduring mystery as to why human chromosomes carried in rodent cell hybrids should be distinguishable from their rodent counterparts within the same cell by G11-banding. On primate chromosomes heterochromatin is stained magenta while euchromatin is blue, whereas rodent chromosomes are the opposite—magenta in euchromatin and blue in heterochromatin.[80] A blue coloration is thought to result from binding of the thiazine components of Giemsa alone, whereas magenta is the color produced by binding thiazines and eosin.[2] Since in a mouse-human hybrid the human DNA is packaged with murine chromosomal proteins it seems most likely that the staining reflects differences in the primary DNA sequence of the two species (chapter 3) or in the interaction of the sequence with proteins.

HOW MANY BANDS?

Soon after the discovery of chromosome banding techniques, a standard banding nomenclature was established.[81] This came in the form of an idealized karyotype, or ideogram of all 24 different chromosomes, with the G-bands portrayed in black and the R-bands in white. The total number of bands, or the "resolution" was approximately 350 and this is thought to correspond to chromosomes late in metaphase. The earlier in mitosis the chromosomes are prepared and the longer the chromosomes are, the number of bands that can be resolved by the trained cytogeneticist is increased. The role of increasing chromosome condensation as mitosis progresses is discussed in chapter 5.

The use of colcemid or other spindle poisons accumulates chromosomes late in metaphase and hence prophase chromosomes in such preparations are rare. There are no drugs that allow prophase chromosomes to be accumulated specifically. Attempts to increase the proportion of long mitotic chromosomes for high-resolution banding often rely on synchronization of cells in the interphase prior to harvest. This can be achieved with any of those agents that interfere with S phase and that can be used for replication banding and whose mode of action we have already discussed. Judicious choice of the time of chromosome harvest after release from the blocking agents can result in increased proportions of cells in prophase and early stages of metaphase but this is at the expense of the mitotic index. Alternatively, agents that generally interfere with chromosome condensation can be used. The most commonly used of these are the intercalating dyes acridine orange and ethidium bromide and we assume that the mere presence of these dyes between the stacked base pairs of the DNA helix acts as a physical barrier to condensation.[82,83] The reduction of disulphide bonds by β-mercaptoethanol also appears to inhibit chromosome condensation.[84]

Ideograms of the human karyotype consisting of approximately 400 or 550 (mid to early metaphase), 850 or 1250 (prometaphase) and 2000 G-bands (late prophase) have been published.[85] High-resolution ideograms (approximately 1250 bands) have also been produced for R-banded chromosomes.[86]

As chromosomes progressively condense during metaphase the high-resolution chromosome bands coalesce, with smaller bands being engulfed by larger ones. There is some evidence that high-resolution R-bands, and particularly T-bands, fuse more often than G-bands.[87] This may reflect real differences in the way that different parts of the genome condense or it may be an optical illusion. Given a haploid human genome content of 3×10^3 Mb, the average 2000 band resolution chromosome band must contain 1.5 Mb of DNA, while the average 450 band resolution band will contain 6-7 Mb of DNA. These estimates give a feel for the resolution limit of cytological analysis of metaphase chromosomes. It is likely that the human genome contains more than 2000 chromosome 'bands' at a molecular level but these domains will be below the cytogenetic resolution limit. Examples of chromosome 11 ideograms at different levels of resolution, for G-bands, R-bands, replication bands and meiotic chromomeres are shown in Figure 2.5.

BANDS IN SHADES OF GREY

The nomenclature of chromosome banding such as G-positive and G-negative bands, together with the portrayal in many chromosome ideograms of bands in either black or white, conveys an image of different types of chromosome bands as opposite extremes with distinct dividing lines between them. In fact, chromosome bands exhibit a range of staining intensities that are not only 'black or white' but all shades of grey in between. Chromosome bands

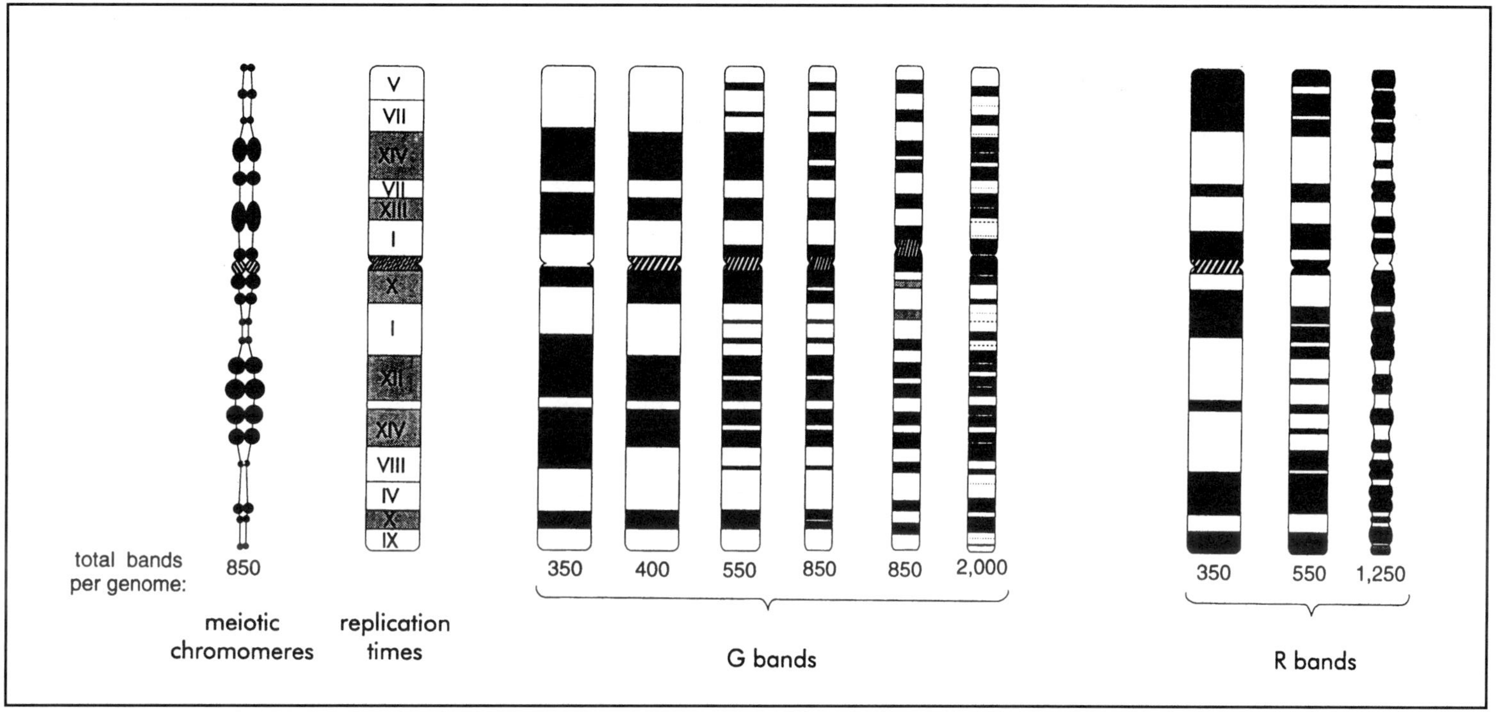

Fig. 2.5 .Ideograms of human chromosome 11 banded in different ways. From left to right are shown; pachytene chromomeres, relative replication times with I being earliest and XIV the latest (from ref. 31), G-bands (black) at increasing levels of resolution from 350 to 2000 bands and lastly, R-bands (black).

have more recently been classified in a more refined way into a variety of 'flavors' that, as we shall see in subsequent chapters, are a reflection of the molecular properties of the DNA and protein in the bands.[20]

At a low-resolution (<300 bands) Q-bands show a variation in brightness.[24] This is also true for the staining intensity of G-bands, with four "shades" of G-band having been recorded. Ideograms based on staining intensity have been updated recently and are based on detailed measurements of actual band sizes and digitalization of the staining intensity (Fig. 2.5).[88] Variations in staining intensity of R-bands have also been noted.[86] Human R-bands have been divided into four "flavors" dependent on a variety of molecular features, ranging from the most extreme T-band flavor R-band to 'mundane' R-bands, which might be considered as the most G-band-like of R-bands.[20]

Replication banding has shown that not only do R-bands replicate before G-bands, but also that different subsets of these bands replicate at different times.[30] S phase has been arbitrarily divided into eighteen time-segments with each G- and R-band assigned a relative replication time.[31] This relative replication time is related to the band flavor and other molecular properties of the chromosome band (chapter 5). The relative replication times for chromosome 11 are shown in Figure 2.5.

Chromosome ideograms portray the boundaries between chromosome bands as distinctive and sharp dividing lines. At a molecular level it is not clear whether there are really sharp transitions from one band to another or a more gradual gradation in properties (chapter 3).

SWELLING AND FIXATION

For practical reasons cells are almost always swollen in hypotonic during the chromosome preparation process. It has been suggested that this in and of itself can induce some structural heterogeneity in the chromosomes and hence contribute to the banding pattern.[89] In general, the use of hypo- and hyper-tonic conditions to study chromosomes has raised some questions as to the physiological relevance of the conclusions.[90] In addition to this, metaphase chromosomes are normally fixed prior to banding, unless antibodies are being used to generate the bands. The most commonly used fixatives for dye-based banding methods are alcohol/acetic acid mixes, typically 3:1 methanol: acetic acid. Acetic acid probably has several rather drastic effects on the chromosome; it is thought to depurinate and nick DNA and to extract histones—especially histone H1[91,92] and it may even change the form of the DNA helix.[66]

Given all of these preconditions, could chromosome banding all be a horrible artifact? We obviously think the answer to this question is no. G-band like-patterns can be seen on the surface of fixed, but otherwise untreated, chromosomes by atomic force microscopy.[93] Banding has also been reported on unfixed chromosomes[94] and chromosomes fixed in formaldehyde also give some DAPI banding. Most definitively, the recapitulation of the cytological banding patterns at a molecular level within the organization of the primary DNA sequence of chromosomes and within the temporal sequence of DNA replication are strong arguments for the case that chromosome bands have physiological relevance.

It is apparent that the principle banding techniques probably exploit differences in the nature of the primary DNA sequence (Q- and perhaps R- and T-banding), chromatin structure (G- and R-banding and meiotic chromomeres) and DNA replication timing (replication banding) along chromosomes from the description of the different ways in which chromosome bands can be elicited. These facets of chromosome structure and organization are those that are explored further in subsequent chapters of this book.

References

1. Verma RS, Babu A. Human Chromosomes: Principles and Techniques, McGraw Hill, 1995.
2. Sumner AT. Chromosome banding. London: Unwin Hyman, 1990.
3. Gosden JR. Chromosome analysis protocols. In: Walker JM, ed. Methods in Molecular Biology. Vol 29. Series Humana Press, 1994.
4. Arrighi FE and Hsu TC. Localization of heterochromatin in human chromosomes. Cytogenet. 1971; 10:81-86.
5. Sumner AT, Evans HJ and Buckland RA. New technique for distinguishing between human chromosomes. Nature New Biology 1971; 232:31-32.
6. Miklos GLG and John B. Heterochromatin and satellite DNA in man: properties and prospects. Am. J. Hum. Genet. 1979;

31:264-280.
7. Singer MF. Highly repeated sequences in mammalian genomes. Int. Rev. Cytol. 1982; 76:67-113.
8. Burkholder GD and Weaver MG. DNA-protein interactions and chromosome banding. Exp. Cell Res. 1977; 110:251-262.
9. Dutrillaux B, de Grouchy J, Finaz C and Lejeune J. Mise en évidence de la structure fine des chromosomes humains par digestion enzymologique (pronase en particulier). Comptes Redus de l'Academie des Sciences (Paris) D. 1971; 276:3179-3181.
10. Seabright M. A rapid banding technique for human chromosomes. Lancet 1971; 2:971-972.
11. Mäkinen T, Stenstrand K and Selander R-K. The effect of ionic strength on G-bands. Humangenetik 1975; 28:71-73.
12. van Duijn P, van Prooijen-Knegt AC and van der Ploeg M. The involvement of nucleosomes in Giemsa staining of chromosomes. Histochemistry 1985; 82:363-376.
13. Harrison CJ, Britch M, Allen TD and Harris R. Scanning electron microscopy of the G-banded human karyotype. Exp. Cell Res. 1981; 134:141-153.
14. Harrison CJ, Jack EM, Allen TD and Harris R. Light and scanning electron microscopy of the same human metaphase chromosomes. J. Cell Sci. 1985; 77:143-153.
15. Dutrillaux B and Lejeune J. Sur une nouvelle technique d'analyze du caryotype humaine. Comptes Rendus de l'Academie des Sciences (Paris) D, 1971; 272:2638-2640.
16. Bobrow M and Madan K. The effects of various banding procedures on human chromosomes, studied with acridine orange. Cytogenet. Cell Genet. 1974; 12:145-156.
17. Bernheim A, Berger R and Szabo P. Localization of actin-related sequences by in situ hybridization to R-banded human chromosomes. Chromosoma 1984; 89:163-167.
18. Dutrillaux B. Nouveau systeme de marquage chromosomique les bandes T. Chromosoma 1973; 41:395-402.
19. Ambros PF and Sumner AT. Correlation of pachytene chromomeres and metaphase bands of human chromosomes, and distinctive properties of telomeric regions. Cytogenet. Cell Genet. 1987; 44:223-228.
20. Holmquist GP. Chromosome bands, their chromatin flavors, and their functional features. Am. J. Hum. Genet. 1992; 51:17-37.
21. Jack EM, Harrison CJ, Allen TD and Harris R. A structural basis for R- and T-banding: a scanning electron microscopy study. Chromosoma 1986; 94:395-402.
22. Caspersson T, Farber S, Foley GE et al. Chemical differentiation along metaphase chromosomes. Exp. Cell Res. 1968; 49:219-222.

23. Caspersson T, Zech L and Johanssen C. Differential binding of alkylating fluorochromes in human chromosomes. Exp. Cell Res. 1970; 60:315-319.
24. Korenberg JR and Engels WR. Base ratio DNA content and quinacrine brightness of human chromosomes. Proc. Natl. Acad. Sci. USA 1978; 75:3382-3386.
25. Ferruci L and Mezzanotte R. A cytological approach to the role of guanine in determining quinacrine fluorescence response in eukaryotic chromosomes. J. Histochem. Cytochem. 1982; 30:1289-1292.
26. Kubista M, Akerman B and Norden B. Characterization of interaction between DNA and 4',6-diamidino-2-phenylindole by optical spectroscopy. Biochem. 1987; 26:4545-4553.
27. Schweizer D. Counterstain-enhanced chromosome banding. Hum. Genet. 1981; 57:1-14.
28. Saitoh Y and Laemmli UK. From the chromosomal loops and the scaffold to the classic bands of metaphase chromosomes. Cold Spring Harbor Symp. Quant. Biol. 1993; 153:755-765.
29. Holmquist G, Gray M, Porter T and Jordan J. Characterization of Giemsa dark and light-band DNA. Cell 1982; 31:121-129.
30. Ganner E and Evans HJ. The relationship between patterns of DNA replication and of quinacrine fluorescence in the human chromosome complement. Chromosoma 1971; 35: 326-341.
31. Dutrillaux B, Couturier J, Richer C-L and Viegas-Péguinot E. Sequence of DNA replication in 277 R- and Q-bands of human chromosomes using a BrdU treatment. Chromosoma 1976; 58:51-61.
32. Vogel W, Autenrieth M and Mehnert K. Analysis of chromosome replication by a BrdU antibody technique. Chromosoma 1989; 98:335-341.
33. Drouin R, Holmquist GP and Richer C-L. High resolution replication bands compared with morphological G- and R- bands. Adv. Hum. Genet. 1994; 22:47-115.
34. Reichard P. Interactions between deoxyribonucleotide and DNA synthesis. Ann. Rev. Biochem. 1988; 57:349-374.
35. Holmquist GP and Comings DE. Sister chromatid exchange and chromosome organization based on a bromodeoxyuridine Giemsa—C-banding technique (TC-banding). Chromosoma 1975; 52:45-259.
36. Klever M, Grond-Ginsbach C, Schertan, H and Schroeder-Kurth TM. Chromosomal in situ suppression hybridization after Giemsa banding. Hum. Genet. 1991; 86:484-486.
37. Fantes JA, Oghene K, Boyle S et al. A high resolution integrated physical, cytogenetic and genetic map of human chromosome 11: distal p13 to proximal p15.1. Genomics 1995; 25:447-461.
38. Korenberg JR and Chen XN. Human cDNA mapping using a high-resolution R-banding technique and fluorescence in situ hybridiza-

tion. Cytogenet. Cell Genet. 1995; 69:196-200.
39. Cherif D, Julier C, Delatte O et al. Simultaneous localization of cosmids and chromosome R-banding by fluorescence microscopy: application to regional mapping of human chromosome 11. Proc. Natl. Acad. Sci. USA 1990; 87:6639-6643.
40. Lemieux N, Dutrillaux B and Viegas-Pequignot E. A simple method for simultaneous R- or G-banding and fluorescence in situ hybridization of small single-copy genes. Cytogenet. Cell Genet. 1992; 59:311-312.
41. Takahashi E, Hori T, O'Connell P et al. R-banding and non-isotopic in situ hybridization: precise localization of the human type II collagen gene (COL2AI) Hum. Genet. 1990; 86:14-16.
42. Okada TA and Comings DE. Mechanisms of chromosome banding. III. Similarities between G-bands of mitotic chromosomes and chromomeres of meiotic chromosomes. Chromosoma 1974; 48:65-71.
43. Jhanwar SC, Burns JP, Alonso ML et al. Mid-pachytene chromomere maps of human autosomes. Cytogenet. Cell Genet. 1982; 33:240-248.
44. Luciani JM, Guichaoua MR, Cau P et al. Differential elongation of autosomal pachytene bivalents related to their DNA content in human spermatocytes. Chromosoma 1988; 97:19-25.
45. Kerem B-S, Goitein R, Diamond G et al. Mapping of DNAase I sensitive regions on mitotic chromosomes. Cell 1984; 38:493-499.
46. Adolph S and Hameister H. In situ nick translation of metaphase chromosomes with biotin-labeled dUTP. Hum. Genet. 1985; 69:117-121.
47. Alfi OS, Donnell GN and Derencsenyi A. C-banding of human chromosomes produced by DNase. Lancet 1973; 1:505.
48. Burkholder GD. Morphological and biochemical effects of endonucleases on isolated mammalian chromosomes in vitro. Chromosoma 1989; 97:347-355.
49. Ferraro M, Predazzi V and Prantera G. In human chromosomes telomeric regions are enriched in CpGs relative to R-bands. Chromosoma 1993; 102:712-717.
50. Sumner AT, Taggart MH, Mezzanotte R and Ferruci L. Patterns of digestion of human chromosomes by restriction endonucleases demonstrated by in situ nick translation. Histochem. J. 1990; 22:639-652.
51. Gosálvez J, Lopez-Fernandez C, Ferrucci L and Mezzanotte R. DNA base sequence is not the only factor for restriction endonuclease activity on metaphase chromosomes: evidence using isoschizomers. Cytogenet. Cell Genet. 1989; 50:142-144.
52. Jeppesen P. Immunofluorescence techniques applied to mitotic chro-

mosome preparations. In: Gosden JR, ed. Methods in Molecular Biology. Vol 29. Chromosome analysis protocols. Humana Press Inc. 1994:253-285.
53. Sullivan BA and Schwartz S. Identification of centromeric antigens in dicentric Robertsonian translocations: CENP-C and CENP-E are necessary components of functional centromeres. Hum. Mol. Genet. 1995; 4:2189-2197.
54. Magaud J-P, Rimokh R, Brochier J et al. Chromosomal R-banding with a monoclonal anti double-stranded DNA antibody. Hum. Genet. 1985; 69:238-242.
55. Miller OJ, Schnedl W, Allen J and Erlanger BF. 5-Methylcytosine localized in mammalian constitutive heterochromatin. Nature 1974; 251:636-637.
56. Lewis JD, Meehan RR, Henzel WJ et al. Purification, sequence, and cellular localization of a novel chromosome protein that binds to methylated DNA. Cell 1992; 69:905-914.
57. Wreggett KA, Hill F, James PS et al. A mammalian homologue of Drosophila heterochromatin protein 1 (HP1) is a component of constitutive heterochromatin. Cytogenet. Cell Genet. 1994; 66:99-103.
58. Disney JE, Johnson KR, Magnuson NS et al. High-Mobility Group protein HMG-I localizes to G/Q- and C-bands of human and mouse chromosomes. J. Cell Biol. 1989; 109:1975-1982.
59. Breneman JW, Yau P, Teplitz RL and Bradbury EM. A light microscope study of linker histone distribution in rat metaphase chromosomes and interphase nuclei. Exp. Cell Res. 1993; 206:16-26.
60. Jeppesen P and Turner BM. The inactive X chromosome in female mammals is distinguished by a lack of histone H4 acetylation, a cytogenetic marker for gene expression. Cell 1993; 74:281-289.
61. Burkholder GD, Latimer LJP and Lee JS. Immunofluorescent staining of mammalian nuclei and chromosomes with a monoclonal antibody to triplex DNA. Chromosoma 1988; 97:185-192.
62. Kapuscinski J and Szer W. Interactions of 4',6-diamidine-2-phenylindole with synthetic polynucleotides. Nucl. Acids Res. 1979; 6:3519-3534.
63. Plumbridge TW and Brown JR. The interaction of adriamycin and adriamycin analogs with nucleic acids in the B and A conformations. Biochemica Biophys. Acta 1979; 563:181-192.
64. Rich A, Nordheim A and Wang AH-J. The chemistry and biology of left-handed Z-DNA. Ann. Rev. Biochem. 1984; 53:791-846.
65. Viegas-Péquinot E, Derbin C, Malfoy, B et al. Z-DNA immunoreactivity in fixed metaphase chromosomes of primates. Proc. Natl. Acad. Sci. USA 1983; 80:5890-5894.
66. Hill RJ and Stollar BD. Dependence of Z-DNA antibody binding

to polytene chromosomes on acid fixation and DNA torsional strain. Nature 1983; 305:338-340.

67. Ueda T, Kato Y and Irie S. Regional differences in immunostainability of isolated metaphase chromosomes of indian Muntjac with anti-Z DNA antibody. Chromosoma 1990; 99:161-168.
68. Korenberg JR and Rykowski MC. Human genome organization. Alu, Lines, and the molecular structure of metaphase chromosome bands. Cell 1988; 53:391-400.
69. Boyle AL, Ballard SG and Ward DC. Differential distribution of long and short interspersed element sequences in the mouse genome: chromosome karyotyping by fluorescence in situ hybridisation. Proc. Natl. Acad. Sci. USA 1990; 87:7757-7761.
70. Lichter P, Ledbetter SA, Ledbetter DH and Ward DC. Fluorescence in situ hybridization with Alu and L1 polymerase chain reaction probes for rapid characterisation of human chromosomes in hybrid cell lines. Proc. Natl. Acad. Sci. U.S.A 1990; 87:6634-6638.
71. Boyle AL, Feltquite DM, Dracopoli NC et al. Rapid physical mapping of cloned DNA on banded mouse chromosomes by fluorescence in situ hybridization. Genomics 1992; 12:106-115.
72. Saccone S, De Sario A, Wiegant J et al. Correlations between isochores and chromosomal bands in the human genome. Proc. Natl. Acad. Sci. USA 1993; 90:11929-11933.
73. Craig JM and Bickmore WA. The distribution of CpG islands in mammalian chromosomes. Nature Genetics 1994; 7:376-382.
74. Rabbits P, Impey H, Hepell-Parton A et al. Chromosome specific paints from a high resolution flow karyotype of the mouse. Nature Genet. 1995; 9:369-375.
75. Speicher MR, Ballard SG and Ward DC. Karyotyping human chromosomes by combinatorial multi-fluor FISH. Nature Genet. 1996; 12:368-375.
76. Schröck E, du Manoir S, Veldman T et al. Multicolor spectral karyotyping of human chromosomes. Science 1996; 273: 494-497.
77. Lengauer C, Speicher MR, Popp S et al. Chromosomal bar codes produced by multicolor fluorescence in situ hybridization with multiple YAC clones and whole chromosome painting probes. Hum. Mol. Genet. 1993; 2:505-512.
78. Bray-Ward P, Menninger J, Lieman J et al. Integration of the cytogenetic, genetic and physical maps of the human genome by FISH mapping of CEPH YAC clones. Genomics 1996; 32:1-14.
79. Friend KK, Dorman BP, Kucherlapati RS and Ruddle FH. Detection of interspecific translocations in mouse-human hybrids by alkaline Giemsa staining. Exp. Cell Res. 1976; 99:31-36.
80. Bobrow M and Cross J. Differential staining of human and mouse chromosomes in inter-specific cell hybrids. Nature 1974; 251:77-79.

81. Paris Conference. Standardisation in human cytogenetics. Birth defects: original article series, Vol 8, No 7, The National Foundation, New York 1972.
82. Matsubara T and Nakagome Y. High-resolution banding by treating cells with acridine orange before fixation. Cytogenet. Cell Genet. 1983; 35:148-151.
83. Ikeuchi T. Inhibitory effect of ethidium bromide on mitotic chromosome condensation and its application to high-resolution chromosome banding Cytogenet. Cell Genet. 1984; 38:56-61.
84. Kao YS, Whang-Peng J and Lee E. A simple, rapid, high-resolution chromosome technique for lymphocytes. Am. J. Clin. Path. 1983; 79:481-483.
85. Yunis JJ. Mid-prophase human chromosomes. The attainment of 2000 bands. Hum. Genet. 1981; 56:293-298.
86. Drouin R and Richer C-L. High-resolution R-banding at the 1250-band level II. Schematic representation and nomenclature of human RBG-banded chromosomes. Genome. 1988; 32:425-439.
87. Drouin R, Lemieux N and Richer C-L. Chromosome condensation from prophase to late metaphase: Relationship to chromosome bands and their replication time. Cytogenet. Cell Genet. 1991; 57:91-99.
88. Francke U. Digitized and differentially shaded human chromosome ideograms for genomic applications. Cytogenet. Cell Genet. 1994; 65:206-219.
89. Zatsepina OV, Polyakov VY and Chentsov YS. Differential decondensation of mitotic chromosomes during hypotonic treatment of living cells as a possible cause of G-banding: an ultrastructural study. Chromosoma 1989; 98:109-116.
90. Cook PR. A chromomeric model for nuclear and chromosome structure. J. Cell Sci. 1995; 108:2927-2935.
91. Hutchison N and Weintraub H. Localization of DNaseI-sensitive sequences to specific regions of interphase nuclei. Cell 1985; 43:471-482.
92. Burkholder GD and Duczek LL. The effect of chromosome banding techniques on the proteins of isolated chromosomes. Chromosoma 1982; 87:425-435.
93. Musio A, Mariani T, Frediani C et al. Longitudinal patterns similar to G-banding in untreated human chromosomes: evidence from atomic force microscopy. Chromosoma 1994; 103:225-229.
94. Buys CHCM, Koerts T and van der Veen AY. Banding of unfixed mitotic chromosomes in suspension after release from human lymphocytes and fibroblasts. Hum. Genet. 1984; 66:361-364.
95. Bickmore WA and Sumner AT. Mammalian chromosome bands—an expression of genome organization. Trend Genet. 1989; 5:144-148.

CHAPTER 3

DNA Sequence and Chromosome Bands

The first level of chromosome organization at which compartmentalization of the genome can occur is within the DNA sequence itself. Our knowledge about this level of genome organization is increasing apace with the rapid progress in efforts to sequence the entire genomes of a number of eukaryotes including at least one mammal (man). Here we consider to what extent variations in gene density, base composition and the distribution of repetitive DNA sequences relate to, and could contribute to, chromosome banding patterns in vertebrates.

DNA METHYLATION AND CpG ISLANDS

Only a few percent of the mammalian genome has a coding function and several lines of evidence indicate that genes are nonrandomly distributed in the genome. Some of the first clues to this resulted from analyzing the hybridization profiles of radioactively labeled human cDNA, poly(A) mRNA and heterogeneous nuclear RNA to human chromosomes—70-80% of the resulting signal resided over R-bands (at a resolution of 300 bands).[1]

The genomes of vertebrates provide us with readily isolated and quantifiable tags on genes that are independent of the level and site of expression of these genes. Vertebrate genomes are methylated at most CpG dinucleotides and in most of the genome CpG occurs at a much lower frequency than GpC. Depletion of CpG is thought to be due to the unrepaired deamination of 5-methylCpG (^{5Me}CpG) and consequent C to T change.[2] The

5' ends of 50-60% of mammalian genes are associated with 0.2-2 kb regions called CpG islands. Although they account for only 1-2% of the genome, CpG islands stand out from bulk genomic DNA because they are generally unmethylated, show no CpG depletion versus GpC (reflecting their unmethylated state) and, at least in mammals, also tend to be GC-rich.[3]

Around 60% of human genes (all housekeeping and ~40% of tissue-specific genes) and 50% of mouse genes possess CpG islands.[4,5] The characteristics of islands means that they are the main cutting sites in mammalian genomic DNA for certain methylation-sensitive restriction enzymes with CpGs in their recognition sites.[6] These enzymes can be used to isolate CpG islands themselves or to cut genomic DNA specifically at island-associated genes and not in the intervening DNA. In the latter case, the size of the resulting fragments is inversely proportional to CpG island density. This is illustrated in Figure 3.1 for a hypothetical stretch of mammalian DNA and in Figure 3.2 for a real piece of the human genome—a 13 kb stretch of the human alpha (α) globin gene complex from chromosome 16p13. Whole CpG islands (CGIs) can also be selected by affinity purification on columns containing a methylated DNA binding protein MeCP2.[7]

The sites in the human genome where most CpG islands are located have been identified by hybridizing CpG islands to human metaphase chromosomes by fluorescence in situ hybridization (FISH).[8,9] The resulting pattern of hybridization signal shows a striking variation in CpG island distribution both between and within the chromosomes of the human karyotype (Fig. 3.3). Most islands appear to reside in the subset of early-replicating, GC-rich R-bands, known to cytogeneticists as T-bands (chapter 2). The T-band karyotype of human chromosomes can be seen in Figure 3.4 and it corresponds very well with the CpG island karyotype (for example, compare chromosomes 1 in Figures 3.3 and 3.4).

Human chromosomes that are generally CpG island-rich throughout their length, as judged from Figure 3.3, include #17, 19 and 22. Thus chromosome 19, where all the R-bands are classified as T-bands, is very CpG island-rich, while the comparably sized chromosome 18, where none of the R-bands are T-bands and which has prominent late replicating G-bands, is island-poor

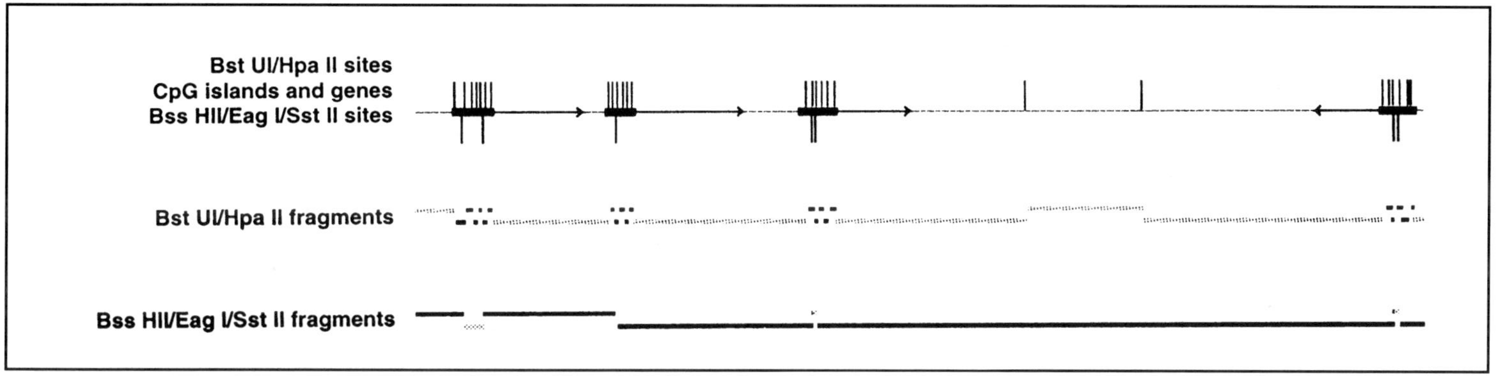

Fig. 3.1. The distribution of sites for CpG island-specific enzymes in a hypothetical stretch of mammalian genomic DNA. CpG islands are shown as black boxes and genes as horizontal arrows. Bst UI (CGCG) and Hpa II (CCGG) digest islands down to small fragments and sites for these enzymes are shown above the line. Bss HII (GCGCGC), Eag I (CGGCCG) and Sac II (CCGCGG) cleave a few times within islands and rarely in between.[6] Sites for these enzymes are shown below the line. The resulting fragments are shown below the main map. Fragments isolated as FISH probes to analyze CpG island distribution (Fig. 3.3) and CpG island density[8] are shown as the filled lines.

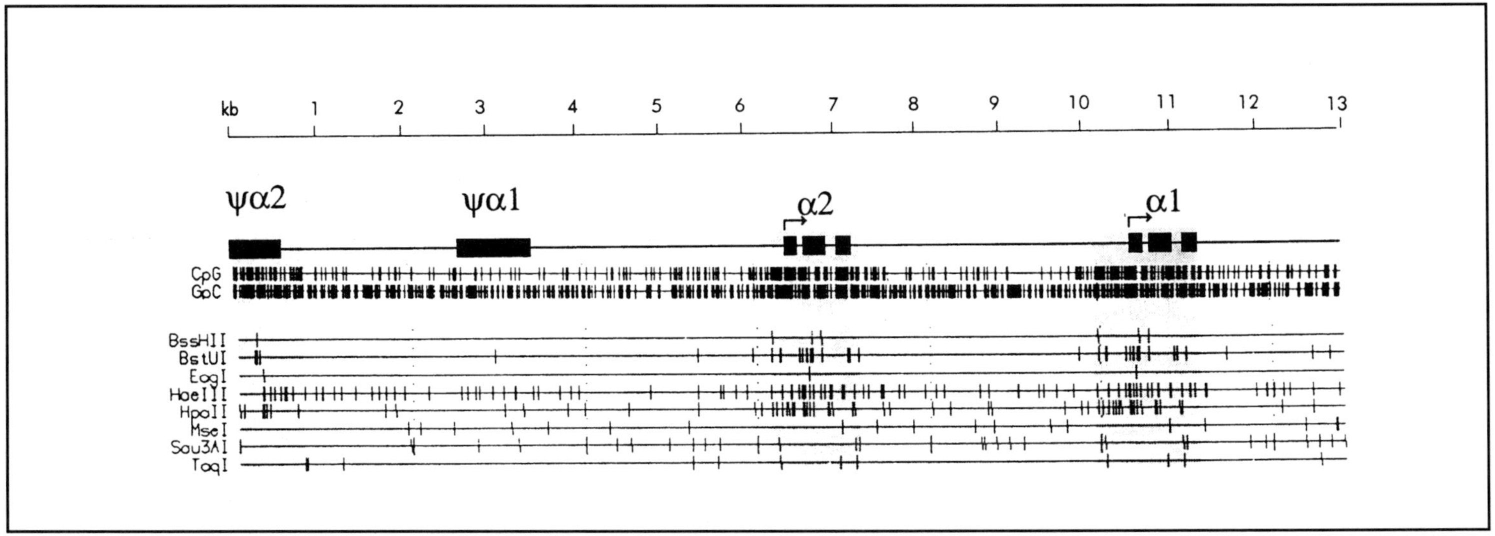

Fig. 3.2. CpG islands in 13 kb of the human α globin complex in the R-band 16p13. Restriction sites and the location of CpG and GpC dinucleotides were taken from the sequence hba4 in Genbank that encompasses the expressed α-globin genes α1 and α2 as well as the two pseudogenes ψα1 and ψα2. The exons of these genes are shown as black boxes and the extent of CpG islands is shaded in grey. Unlike most pseudogenes, including ψα1, ψα2 has yet to lose its CpG island.[84] Note the clustering of sites for the methylation-sensitive enzymes with CpG-containing recognition sequences (Bss HII, Bst UI, Eag I, Hpa II) within the CpG islands and the tendency for Mse I sites (TTAA) to be outside of islands.

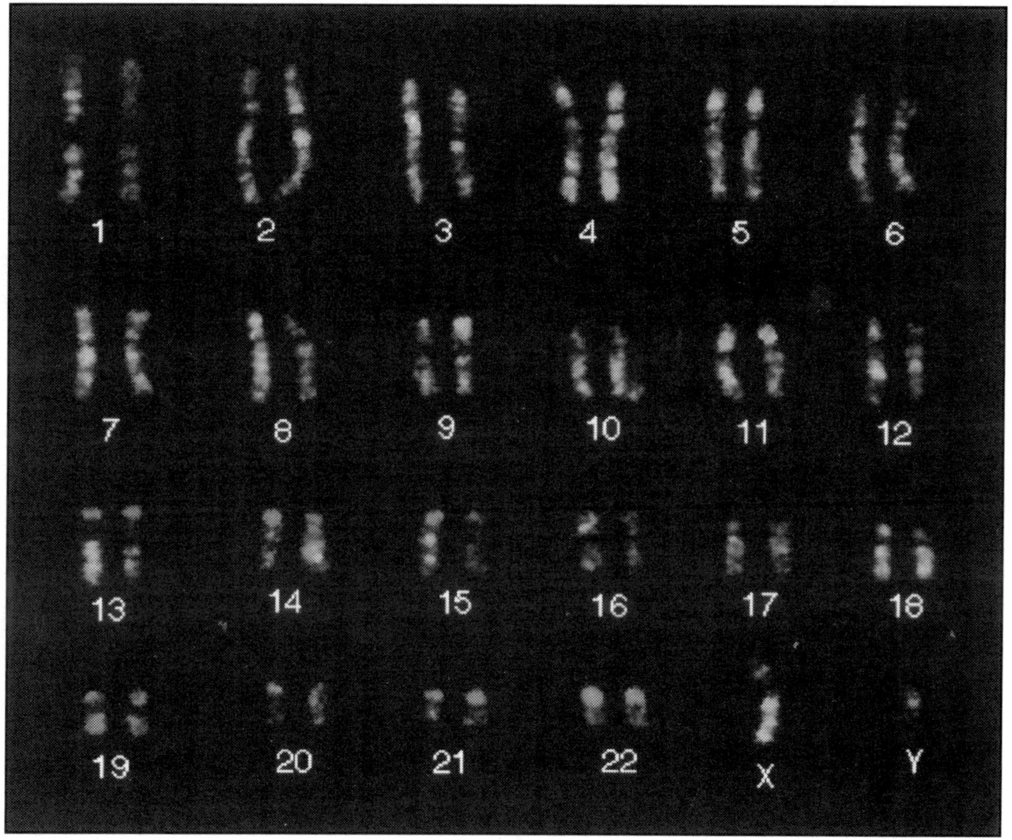

Fig. 3.3. Distribution of CpG islands in the human karyotype. CpG islands were purified as small BstU1 fragments (Figs. 3.1 and 3.2), biotin-labeled and hybridized to human metaphase chromosomes that had been replication banded with BrdU as described in reference 8. CpG islands are visualized in red. Late-replicating G-bands were detected with anti-BrdU-FITC (green). See color figure in insert.

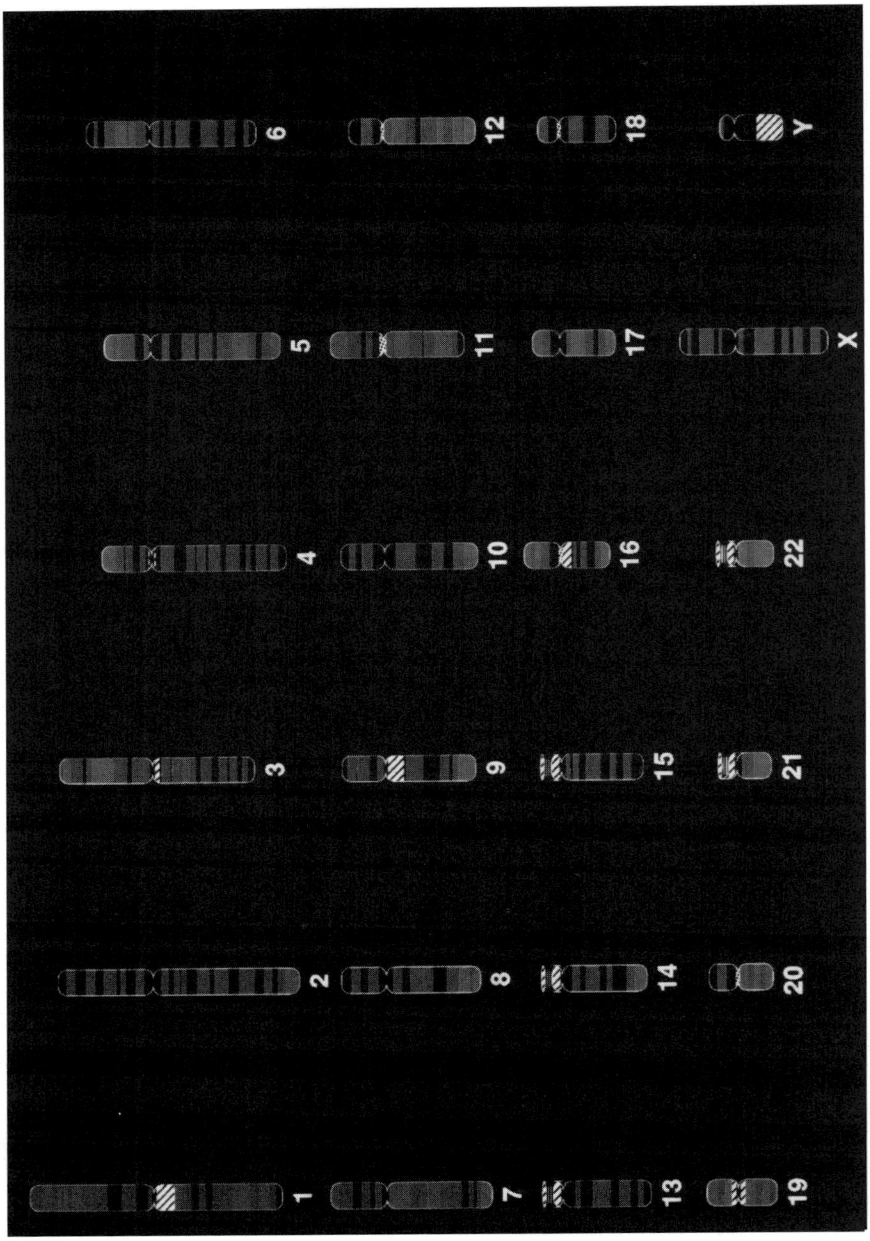

Fig. 3.4. Human chromosome ideogram. G-bands are shown in green, T-bands in red and other R-bands in black, at a resolution of 400 bands per genome. Adapted from ref. 12. See color figure in insert.

(Fig. 3.3). Chromosomes that are CpG-island poor overall include, in addition to chromosome 18, chromosome 13, and to a lesser extent, chromosome 21. The prominent hybridization signals that can be seen on the short arms of the acrocentric chromosomes, including chromosomes 13 and 21, are from the rDNA repeats that are located there and that are known to contain unmethylated GC-rich DNA.[10]

A similar sort of analysis, not with CpG islands themselves but with fractions of inter-island DNA of differing size ranges, showed that T-bands are not anywhere where most human CpG islands are found but also where they are most densely packed. Most of the islands in the human genome seem to be located within 100 kb of each other.[8] DNA that has a spacing of between 100 and 500 kb between islands also appears to be found principally in T-bands, but some of it occurs in many non-T R-bands of the human karyotype. Other R-bands have a CpG island density that is slightly lower than this. The DNA of the human genome that is characterized by very large distances between islands (>1000 kb) was found to correspond to late replicating G-bands and centromeric heterochromatin.[11] It is estimated that <20% of islands are separated from each other by >1000 kb, even though this DNA makes up over half of the human genome.[8]

These experiments revealed that there is a striking relationship between a fundamental feature of the primary DNA sequence—the CpG island density—and not just the chromosome band type but also the more subtle band flavor revealed in the metaphase chromosome by classical banding techniques.[12]

Analysis of long-range restriction maps of the genome also confirms these trends. A complete Not I restriction map has been completed of the long arm of chromosome 21,[13] albeit with the caveat that the map was constructed from a somatic cell hybrid so that some Not I sites may have become methylated in culture and hence not cut. Not I is an enzyme that cuts almost exclusively within CpG islands and will cut at approximately one third of the islands in the human genome.[6] In the proximal 16.7 Mb of 21q that is comprised, in the main, of the large G-bands q21.1 and q21.3, there were nine Not I sites with an average fragment size of 2.1 Mb. In the distal 24.8 Mb, that cytogenetically are dominated by the R-bands q22.1 and q22.3, there are 35 Not I sites with an average spacing of 0.7 Mb. This is consistent with the results of CpG island painting on human chromosome 21.[8,9]

A similar clustering of CpG islands into R-bands is seen in rodent genomes.[8,14] FISH analysis in the mouse has shown that, as in man, CpG islands are the most concentrated in the early-replicating R-band compartments of the karyotype. Almost the entire length of the CpG island-rich chromosome HSA17 is in a conserved synteny group with 45 cM of the distal half of MMU11. Similarly, HSA19q is syntenic with the proximal third (22 cM) of MMU7. Both of these regions of the mouse karyotype hybridize fairly strongly with mouse CGIs and they replicate in the early half of S phase.[15] Conversely, the late-replicating CpG island-poor HSA13q has blocks of conserved synteny with the distal half of MMU14—a region of the mouse karyotype that is CpG island-poor by FISH and late replicating. Hence, there has been a conservation of chromosome organization at the chromosome band level in terms of relative CpG island density, time of DNA replication and banding characteristics over 100 million years of mammalian chromosome evolution. This suggests either the influence of a strong selective pressure to maintain these characteristics together in particular chromosome bands or the action of common mechanisms linking the properties of gene density, replication time and band type functionally.

Despite the similarities in band characteristics between man and mouse the gross organization of the karyotype in the two mammals is quite different. Both the inter- and intrachromosomal variation in CpG island distribution in the mouse is less extreme than that found in the human genome. No autosome in the mouse karyotype appears to show a concentration of CpG islands as high as that of HSA17 or 19 or as low as that of HSA13 or 18.[14] This may be a molecular reflection of the generally less distinct nature of chromosome bands in the mouse. It seems likely that the gross organization of an ancestral vertebrate genome was in fact more like that seen in the human genome than in the mouse, i.e., with entire chromosomes or large blocks of them either very gene-rich or very gene-poor. Chromosomes of a nonmammalian vertebrate (the chicken) show a striking concentration of CpG islands in distinctive parts of the karyotype—the microchromosomes—that comprise only 25% of the genome, rather than on the macrochromosomes.[16] The euchromatic parts of some of these microchromosomes are GC-rich and replicate earlier than any part of the larger chromosomes.[17,18] It has been suggested that the genomes of rodents have become more scrambled than those of other mammals as compared to the genome of a common vertebrate ancestor.[14]

In animals, classical CpG islands are confined to vertebrates, but the genomes of many higher plants are heavily methylated at both CpG and CpXpG residues. Clusters of unmethylated CpGs can be found in these plant genomes and, although they may not be completely analogous to vertebrate CpG islands and must have evolved independently, they do also appear to be associated with genes.[19] As in mammals, restriction enzymes such as Not I have been used to detect such sites on plant chromosomes. These studies have indicated that the plant equivalents of CpG islands are clustered toward the telomeric regions and away from the pericentromeric regions of chromosomes from the major crop species.[20]

GENE DISTRIBUTION AND CHROMOSOME BANDS

We have equated CpG island density with gene density in the discussion so far. Is this justified and can CpG island clustering be extrapolated to gene density itself given that islands are only associated with a subset of all genes? The answer appears to be yes in both mouse and man. Analysis of CpG island distribution in the human genome suggests that >80% of CpG islands are in the 45% of the genome that corresponds to R-bands.[8] So far there is no evidence for gross segregation of genes with and without islands in the genome. By January 1995 a total of 1771 genes had been located to single chromosome bands. This is approximately 2% of the estimated 80,000 human genes,[5] though we can not be sure how representative this sample is of the total. Of these 1771 genes, 354 (20% of the total) map to G-bands, 1417 (80%) to R-bands, and 808 (57%) to T-bands (Fig 3.5). Housekeeping genes are found almost exclusively in R-bands. Of the 230 mapped genes that are fully sequenced at their 5' ends, 90% of genes with CpG islands and 77% of the genes without islands mapped to R-bands

(Table 3.1).[4] At the end of 1995 there were 29 genes regionally assigned on the long arm of human chromosome 21, all but three of which were in the distal q22 region. This indicates an even more striking concentration of genes to the R-band chromatin than was apparent from the Not I map.[13] Numerous random expressed sequences (ESTs) have also been regionally mapped in the human genome, and their overall distribution also shows a concentration into R-band chromatin.[21] On human chromosome 21 there were 97 regionally assigned ESTs in the databases at the time of writing—87% of these were in q22. Recently, direct selection of cDNAs for specific chromosomes has been performed. FISH experiments with the resulting clones showed very similar profiles to those obtained with CpG island paints on the corresponding chromosomes.[22]

The distribution of genes in the mouse genome also mirrors the FISH profile obtained with mouse CpG islands.[14] Of the 253 genes mapped to MMU14 in the Mouse Genome Database (MGD) 208 (82%) are located in the proximal 25 cM of the chromosome that corresponds to an early replicating CpG island rich region, whereas the distal 25 cM of the chromosome (late-replicating and CpG island-poor) contains only 45 (18%).[14,15] As was the case in man, FISH with cDNAs selected for particular mouse chromosomes reinforces the evidence in favor of a relative concentration of genes into the R-band compartment of the rodent genome.[23]

Table 3.1. The location of genes grouped according to function and the presence of a CpG island

Gene Type	% of Genes in R Bands	n
Housekeeping	96%	56
All with CpG islands	90%	131
Tissue-specific with CpG islands	85%	75
All genes	**84%**	**230**
Tissue-specific (all)	80%	173
Tissue-specific without CpG islands	77%	98

All types of gene are located preferentially in R-bands, although genes associated with CpG islands are more likely to be found in R-bands. n is the total number of genes in that particular category. Genes least likely to be found in R-bands are tissue-specific genes not associated with CpG islands.

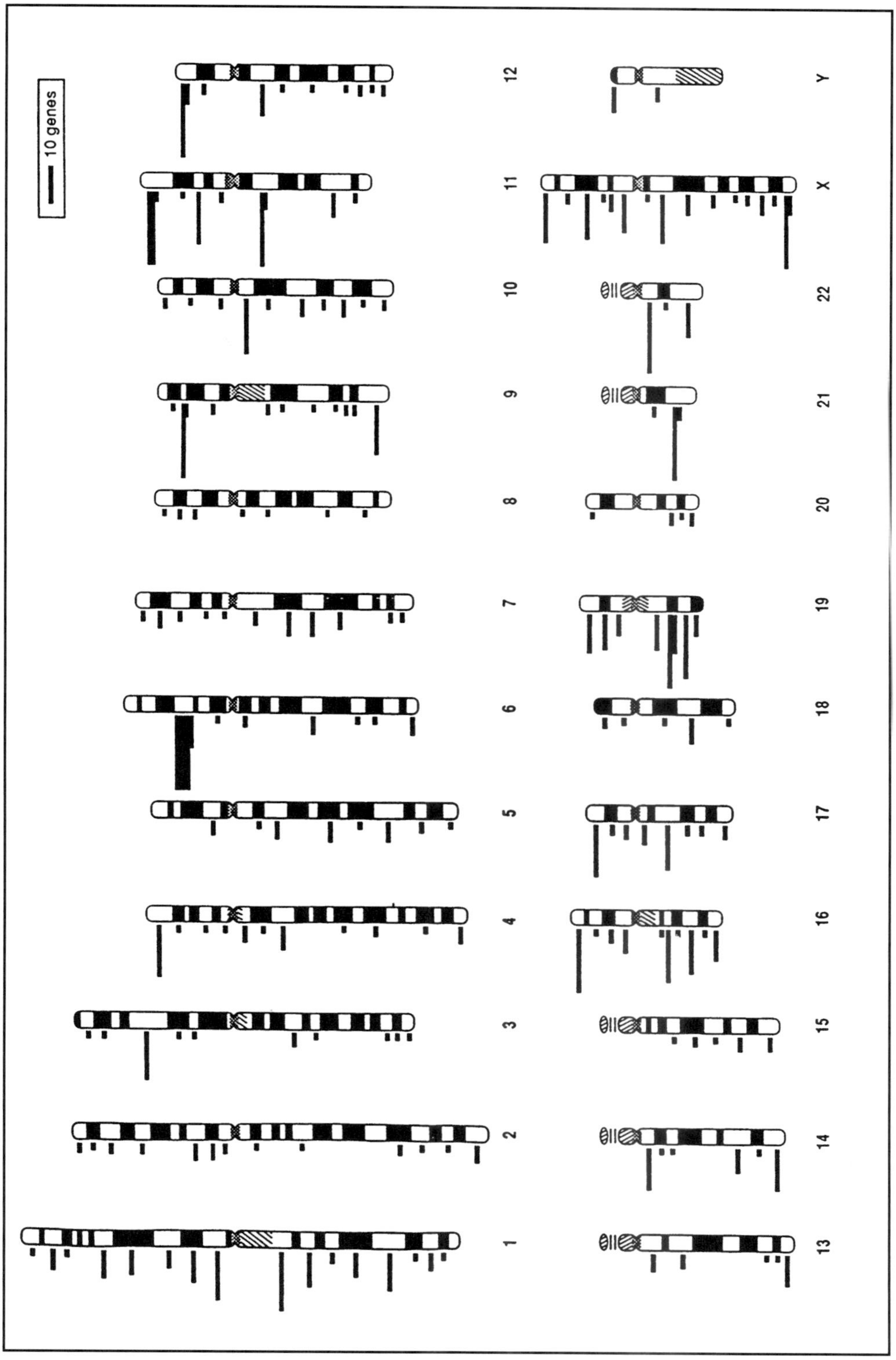

Fig. 3.5. Gene distribution in the human karyotype. Histograms of 1771 genes that have been mapped to single chromosome bands at a resolution of 400 bands per genome in GDB. The ideograms show C-positive bands in black, T-bands and other R-bands are white. The bar represents 10 genes.

The assignment of genes to different chromosomes in the human karyotype is completely consistent with the picture obtained by FISH with CpG islands. The number of gene assignments to the small chromosome HSA19 (7% of the total) is second only to that for HSA1 and the X chromosome (Fig. 3.6). The three autosomes with the smallest number of gene assignments in GDB are HSA18 (<1% of the total) and 21 and 13 (each 1.5% of the total). These three autosomes are also the only ones for which trisomies commonly survive to term in man—reflecting at a clinical level their relatively low gene loads.[24] For gene assignments to mouse chromosomes it is the smaller chromosomes of the karyotype that have the lowest number of genes allocated to them.[14] Again, this is recapitulated in the mouse autosomal trisomies that are born—these are trisomies for the smallest autosomes from MMU13 to 19, excepting MMU15 and 17.[25]

CpG island distribution seems to be a very good guide to the distribution of all genes in mammals. Is this also true of the other vertebrate where we know cytogenetically where the CpG islands cluster—the chicken? Unfortunately it is too early to give the answer to this although there seems no reason to suppose that it should be otherwise. Technical limitations have strongly biased the gene localizations that have been made so far in the chicken genome.[16]

Concentrations of genes in cytologically distinctive chromosome domains is not even restricted to animals. The analysis of unmethylated CpG clusters in plants showed them to be concentrated to the more distal regions of the chromosomes, that are C-band (G-band) negative. It is expected that this will also reflect the distribution of genes on these chromosomes and it is reminiscent of the tendency of T-bands (the most gene-rich) to be located telomerically in the human karyotype.[20,26] We do not know whether this preference for telomeric locations of genes in man is significant or coincidental. The fact that the regions of conserved synteny for genes that in man are at telomeric T-bands, are in the mouse most often located interstitially, argues in favor of the latter. For example, the human α-globin locus is located in a terminal T-band only a few hundred kilobases away from the actual

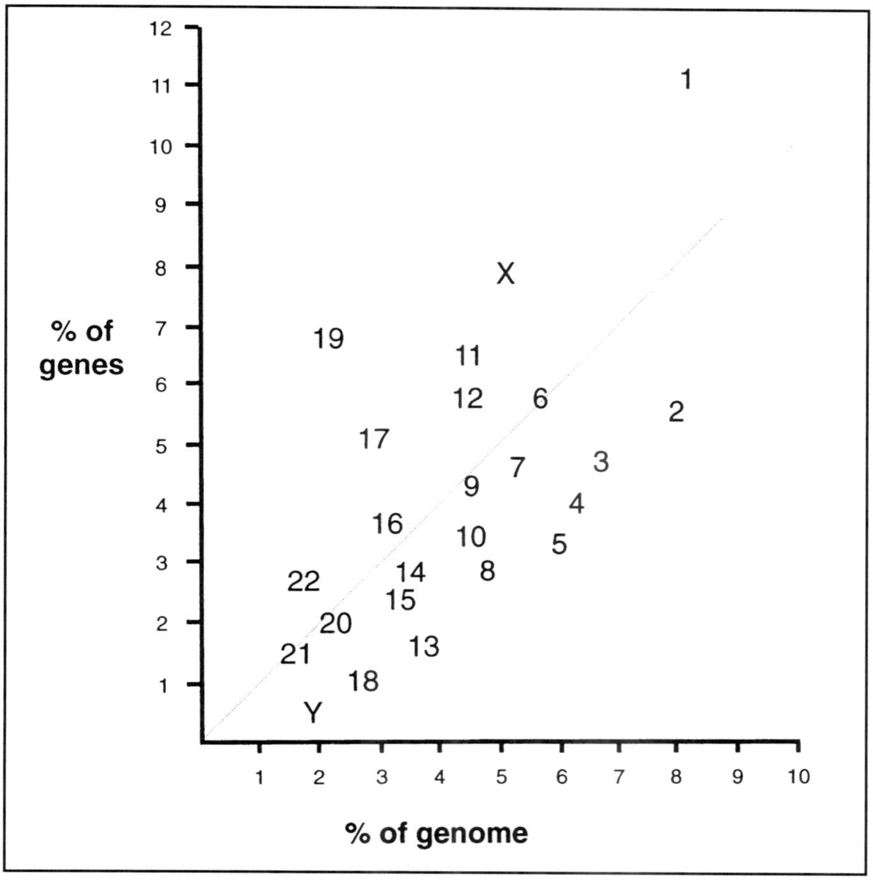

Fig. 3.6. Genes located on human chromosomes in relation to chromosome length. The number of genes localized to each human chromosome in GDB as of January 1996 as a percent of the total is shown relative to the physical proportion of the genome occupied by that chromosome as determined in ref. 85. Note that chromosomes 1 and 19 stand out as being particularly gene-rich chromosomes.

telomere of 16p. In the mouse this locus appears to have become split with the DNA that has conserved synteny with the distal part of human 16p and the distal end of the α-globin locus now located in a proximal location on mouse, about 15 cM from the centromere.[27] Hence there does not seem to be a requirement for the distal human 16p genes to be located close to a functional telomere.

BASE COMPOSITION

Equilibrium centrifugation in analytical cesium salt density gradients has shown that vertebrate and plant genomes are heterogeneous with respect to their base composition. Fractions of DNA with differing base compositions can be prepared by equilibrium centrifugation in Cs_2SO_4 density gradients in the presence of sequence-specific DNA ligands.[28] These fractions have been grouped into a number of families, and the DNA sequences within a group or 'isochore' have similar base compositions and each isochore has its own modal density and base composition. Nonsatellite fractions of human DNA were grouped initially into two GC-poor isochores, L1 and L2, representing approximately 62% of the genome, and three GC-rich isochores, H1, H2 and H3, representing the remaining DNA. The dividing lines drawn between these groups is rather arbitrary. It has been suggested that genomes are made up of these isochores of compositionally homogeneous DNA segments that extend for >300 kb. In support of this, the GC levels of introns, exons and the 3rd codon position of genes seems to correlate well with the overall base composition of the isochore families in which they are embedded.[28,29] More extensive sequence analysis of genomic sequence within different isochore groups will confirm to what extent intergenic DNA obeys the same trend.

In humans, the GC-richest isochore H3 accounts for only approximately 3% of the genome, however, it contains the highest concentration of genes, although there are some genes in all fractions.[29] The chromosomal distribution of the different isochores has been investigated in the human genome by FISH. G-bands hybridize only with the GC-poor isochores (L1 and L2) and R-bands are lit up predominantly by the GC-rich isochores (H1-H3) but also to some extent by the GC-poor isochores. T-bands were identified by the GC-richest (H3) isochore only, confirming their status as the GC-richest bands.[30] As expected from their base composition, rDNA repeats on the acrocentric chromosomes were also sites of hybridization to the H3 isochore.

The similarity between the CpG island and H3 isochore FISH profiles on human chromosomes is further confirmation that the highest gene densities in the genome occur in regions that are also characterized by high GC base compositions, at least in man. The same may not be true of other organisms. The genomes of some crop plants are also characterized by isochores of widely varying base composition and almost all of the genes seem to be concentrated into fragments with a very narrow range of GC content that comprises <20% of the genome. However, this fraction is not the most GC-rich in the genome, rather it seems to occupy the modal position of buoyant density.[31]

Of course, FISH analysis to metaphase chromosomes can only reveal gross band characteristics. Molecular analyses on a finer scale reveal that, although there is a general tendency for G-bands to be quite AT-rich, R bands can contain both AT-rich (L1 and L2) isochores and GC rich isochores (H1, H2 and H3). Thus, sequences within a single chromosome band can have widely-differing GC levels and adjacent R- and G-bands may contain regions with a similar base composition.[32,33] A sharp boundary between different isochores has been found between the human HLA class II (~40% GC) and III (~50% GC) genes in man and the same study also suggested that regions of relatively homogeneous base composition may extend over as much as a megabase of DNA, i.e., into the cytogenetically visible range.[34] Sequence analysis of the boundary identifies tight clusters of LINE and SINE interspersed repeats as well as a sequence with homology to the pseudoautosomal boundary of the sex chromosomes—another place in the genome that marks out chromosome domains which have quite different gene densities, base composition and recombinational behavior from each other. However, this isochore boundary does not appear to correspond to a cytogenetically defined band boundary—the entire MHC region is considered to map in the T-band 6p21.31.[35] Nor does there seem a direct relationship with gene density—the gene and CpG island density in the Class II region appears to be at least as high as that in Class III, consistent with a location for the entire complex within 6p21.3[36]

A striking finding of the analyses on the human chromosomes 21 and X is that GC-content rises dramatically towards the telomere. This might be explained by the terminal location of T-bands in the human genome and their accompanying high gene density. Alternatively, it may be a facet of telomere proximity itself, irrespective of gene density. It will be important to study the base composition of the terminal regions of, for example, mouse chromosomes and of the regions of conserved synteny with human T-bands that are in interstitial locations in the mouse. It has recently been suggested that GC content, at least within the third codon position of coding regions, may be influenced by rates of recombination and/or repair mechanisms.[37] However, more careful analysis of the terminal 8 Mb of human Xq28 reveals a more complex picture. This terminal, but non T- single R-band (even at high resolution banding) comprises a proximal region of 43% GC (H1 isochore), a central region of 47% GC (H2 and H3) and a distal region of only 39.5% GC (L). Hence, a single chromosome band can contain DNA from multiple isochore families but within the band the region of highest GC% still corresponds to the region of highest CpG island and gene density.[38]

How did isochores arise? It has been suggested that isochores evolved in warm-blooded vertebrates to enhance the thermal stability of the gene rich portions of the genome.[28] However, it is hard to imagine that subtle changes in base composition would exert a strong enough selective pressure. It may be equally likely that the existence of isochores in some organisms is just a passive reporter of some other facet of the compartmentalization of the genome. The ways in which base composition might be maintained at a fairly homogeneous level over long stretches of the genome are discussed further in chapter 5.

To what extent can variations in base composition differences contribute to banding patterns? Although variations in base composition in man are sufficient to account for quinacrine banding patterns, they cannot account for those produced by all base-specific fluorochromes.[39] For example, daunomycin will only fluoresce significantly when added to DNA of <35% GC-content. This figure is lower than that of the GC-poorest isochore, and, considering that only a 3.2% difference in base composition has been found between average Q(G)- and R-bands, it seems that simple differences in bulk GC levels alone cannot account for chromosome banding with a fluorochrome such as this.[40] An ingenious alternative explanation to explain chromosome banding with some fluorochromes has recently been put forward[41] and is further discussed in chapter 4.

TANDEMLY REPEATED DNA SEQUENCES

Much of the human genome, and that of many other organisms, comprises arrays of repetitive sequences repeated in tandem. These are termed satellite repeats as they were first identified as satellite peaks on the buoyant density profiles of genomic DNAs. Classical satellite DNA falls into a number of families, each of which is characterized by repetition of a relatively long monomer (of the order of a few hundred base pairs) over many megabases of DNA. The location of satellites on chromosomes often corresponds to the constitutively heterochromatic and nuclease resistant C-bands of mitotic chromosomes and it can also form morphologically distinctive features on meiotic chromosomes.[42] We do not know why satellite DNA should form heterochromatin though it may be related to its very repetitive nature (chapter 6). In man classical satellites I, II and III locate to the pericentromeric C-bands of chromosomes 1, 9, 16 and the long arm of the Y.[43] Satellite DNA is also located on the acrocentric chromosomes just proximal to the rDNA loci. Satellite DNA sequences are assumed to have no function, however, in man and mouse specific satellites (alphoid in man and minor in mouse) are intimately associated with an important functional site on the chromosome—the centromere. This relationship is further discussed in chapter 6.[44]

Tandem repeats based on smaller sized monomers are termed minisatellites (tens of base pairs) and microsatellites (a few base pairs). The polymorphisms between individuals in the size of these arrays has endowed these types of satellite with a central role in human genetics—they are used in genetic fingerprinting and genetic linkage analyses. Unlike the classical satellites, the locations of mini- and micro-satellites do not correspond to cytologically distinctive sites on the chromosome. These repeat arrays do not extend over a long enough stretch of DNA to be visible at mitosis nor, per se, do they seem to cluster into particular flavors of chromosome band. However, it seems that the minisatellites in man that are closest to telomeres (and hence often in T-bands) are those that show the most variability. This indicates that whatever process is responsible for the generation of new variants in these repeat arrays it occurs differentially at different chromosomal sites.[45] It is not clear whether the same is true of minisatellites located in nonterminally located T-bands; many of the hypervariable minisatellites in the mouse genome map to interstitial chromosomal sites.

INTERSPERSED REPEATS

There are two main groups of interspersed repetitive sequences in the mammalian genome, categorized according to their size. Short Interspersed Elements (SINEs) are less than 500 base pairs long and occur 10^5-10^6 times per genome. Long Interspersed Elements (LINEs) are several kilobases long and occur on the order of 10^4-10^5 times.[46] By far the largest members of each of these families in the human genome are the Alu family of SINE and the L1 family of LINE sequences. Alu sequences occur, on average, every 3 kb and occupy approximately 9% of the genome. Full, and 5'-truncated L1s are spaced 150-750 kb and 28 kb apart, respectively, and account for approximately 7% of the genome.[47,48]

Alu and L1 elements differ significantly in their base composition. Alus are 56% G+C, L1s have a GC content of 42%. Hence both of these repeat families are more GC-rich than bulk genomic DNA (40%). In the first uses of FISH to analyze the distribution of classes of DNA sequences across human chromosomes, Alus and LIs were suggested to have grossly reciprocal distributions on human metaphase spreads.[49,50] The Alu hybridization pattern was almost identical to R-banding, with the brightest regions of Alu hybridization corresponding to T-bands, and the least brightest signals corresponding to G-bands. The L1 pattern, though less clear cut, showed strong hybridization to G-bands and to a subset of R-bands.

A similar distribution has also been reported for mouse SINEs (B1 and especially B2) and LINEs by FISH,[51] though there were a few exceptions noted where early replicating, CpG island-rich R-band regions of the mouse karyotype lit up strongly with LINEs. For example, despite the fact that the proximal bands A and B of MMU7 are early replicating,[15] CpG-island rich and contain 224 out of the 347 genes mapped to MMU7[14], they appear to have a high L1 content.[52] The syntenic region to this in man is HSA19q, a region of the human genome that as well as being early replicating and very rich in CpG islands and genes, has a concentration of SINE repeats rather than LINEs as expected.[8,50]

Concomitant with this distribution of interspersed repeats, the trend in the base composition of mouse and human SINEs and LINEs correlates with the base composition of the isochore into which they are embedded, and LINEs and SINEs appear clustered into fragments of different size ranges when genomic DNA from mammals is cleaved with rare-cutter restriction enzymes.[53,54]

However, this simple story does not seem so clear cut by FISH analysis in other mammalian species,[54] nor once you move away from gross cytological studies to high-resolution molecular studies of the human genome, including large scale sequencing. Quite clearly both SINE and LINE repeats can be found in close proximity to each other[55,56] and even over longer distances any differences in distribution between bands types can be hard to discern.[57] Table 3.2 compares the information on gene, GC and Alu content for nine regions of human chromosomes that cytogenetically come from T-, R- and G-bands. Within this limited sample, the highest and lowest Alu densities occur in sequences within a T-band and G-band, respectively. However, there appears to be no strict correlation between Alu density and gene density. There is a positive correlation between Alu density and local GC content, but it seems unlikely that SINE and LINE distributions contribute sufficiently to differences between R- and G-bands, for example, in terms of their influence on base composition, to account for forms of chromosome banding. The contrasting findings on SINE distribution at the molecular versus the cytological level can be reconciled by the suggestion that the SINEs located in R-bands are those that have retroposed most recently and hence are closest in sequence to the progenitor copy.[58] These copies of the repeat will hybridize better to the SINE probes that are based on the consensus repeat sequence than those that are more diverged. Hence copy for copy the R-band SINEs will produce stronger FISH signals than their G-band counterparts.

Why should SINEs retropose preferentially into GC-rich, gene-rich R-bands? It may be a reflection of the different chromatin states of G- and R-bands, which is discussed in subsequent chapters.

Table 3.2. Comparison of gene, GC and Alu content for sequenced regions of the human genome

Locus	Location	Length (kb)	No. Genes	Genes /100 kb	No. of Alus	Alus /kb	% GC
HBB[78]	11p15.4 (G)	88	6	6.8	12	0.14	40
RB1[79]	13q14.3 (R)	180	1	0.6	46	0.26	37
RCP/G6PD[80]	Xq28 (R)	219.5	19	8.7	187	0.85	57
ERCC1[81]	19q13.3 (T)	106	5	4.7	146	1.4	52.5
HLAIII[82]	6p21.3 (T)	80	5	6.3	77	0.96	55
HBA[83]	16p13.3 (T)	54	7	13.0	18	0.33	61

GC-content was taken from ref. 77 or from the mapping reference.

Alternatively, Alus may retropose and integrate in early S phase when R-band chromatin is being replicated whereas L1 movement might be restricted to late S. Thus interspersed repeat distribution would reflect aspects of the control of DNA replication in the germ line. The fact that LINEs can concentrate in early-replicating compartments of mitotic chromosomes might be indicative of differences in the patterns of chromosomal replication between cells of the soma and of the germ line.[51] LINEs seem to be unusual in their genomic distribution. Most dispersed repeats in a wide variety of organisms appear to have a preference for integration into gene rich, GC-rich, chromatin.[59-61]

Is there any functional significance to the distribution of SINEs and LINEs in mammalian genomes? Some transposable elements may perform particular genomic functions. For, example, the LINE-related HET-A and the TART transposable elements are found telomerically and subtelomerically, respectively, on *Drosophila* chromosomes.[62] HET-A elements frequently transpose to broken chromosome ends, and it is thought that they serve to protect the ends of linear DNA molecules in a similar way to the telomeric repeats added by the enzyme telomerase in most other eukaryotes. However, no obvious functional role has been ascribed for mammalian LINEs and SINEs despite several suggestions, but their retroposition certainly has mutational consequences for the genome.[48,63-65]

MODIFIED BASES

Prokaryotic genomes employ a variety of modified bases in their genomes. The predominant modified base in eukaryotic genomes is 5 methylcytosine (5-MeC). This occurs in the context of CpG dinucleotides in vertebrate genomes and both CpG and CpXpG in plants. It may well be the case that there is also some CpXpG methylation in the mammalian genome but its functional significance is unclear.[66] The widespread methylation of CpG dinucleotides in vertebrate genomes has led to a depletion of this dinucleotide during vertebrate evolution, except at CpG islands. Nevertheless, because CpG islands are small (<2 kb) relative to the complete genomic extent of their associated genes, even in very gene-rich regions of the genome such as T-bands, a majority of CpGs will still be methylated. The distribution of 5-MeC has been examined in the mammalian genome by two methods. Firstly, an antibody against 5-MeC demonstrates that this modified base is found at highest concentrations in constitutive heterochromatin of mammalian chromosomes.[67] The same conclusion was arrived at by studying the chromosomal distribution of MECP2—a protein that binds to 5-MeCpG.[68] Methylation appears to play some role in the condensation of these blocks of heterochromatin and chromosome stability since exposure of mammalian cells to the demethylating agent 5-azacytidine results in an unwinding of some sites and reduced levels of 5-MeC in plants gives rise to chromosome instability.[69,70]

G-BAND DNA

What is the nature and function (if any) of the apparently gene-sparse DNA that makes up half of the human genome? There is some evidence that G-band, island-poor DNA may contain local moderately repeated sequences that evolve rapidly.[71] In an attempt to make direct comparisons between human G- and R-band DNAs, examples of these two types of compartment were microdissected and analyzed.[58] Plenty of single copy DNA was found among the G-band clones and no significant differences in base composition could be seen between the two classes of DNA. LINEs were more abundant in the G-band clones, but Alus were no more numerous in the R-band clones than in those from the

G-bands. However, the R-band clones appeared to be concentrated in Alus whose sequence suggested that they had retrotransposed relatively recently. This can explain a lot of the discrepancies between the global and high resolution analyses of SINE distribution.

Some vertebrates, e.g., the pufferfish *Fugu rubripes*, have much smaller intergenic and intronic DNA than mouse or man.[72] It is not known whether this represents a more ancestral state for the vertebrate genome or whether such an organism has more recently disposed of much of its surplus DNA that serves no function. It will be informative to determine whether the missing pufferfish DNA was taken in equal measure from G- and R-band regions of the genome, or whether, because of the longer tracts of intergenic DNA in the former[8] the lion's share of the deficit came from G-band regions. Unfortunately, chromosome banding patterns have not been carried out on pufferfish chromosomes.

Although it now seems clear that, at least in vertebrates, most genes are concentrated into early replicating R-band compartments of the genome, cytological G-bands are by no means completely gene free. Is there anything special about genes that are in G-bands? In general these genes lack CpG islands and are expressed in either temporally or spatially restricted parts of the organism. They are hence often under the control of strong promoter/enhancer elements that recruit a panoply of transcription factors and associated proteins to the job of transcription. Such genes may have evolved to cope with their chromatin environment and might even take advantage of it to aid in their tight regulation to ensure that any inappropriate gene expression is suppressed.[73] Relocating such genes to T-band chromatin might be just as deleterious to the organism as putting some housekeeping genes into G-bands.

Another trend that seems apparent is that genes located in GC-poor isochore fractions tend to encode proteins that are longer than those encoded by genes from the H3 isochore and that the introns are larger in these genes as well.[74] The reason for this trend is not known.

THE CONSEQUENCES OF PRIMARY SEQUENCE ORGANIZATION

This chapter has shown that we need look no further than the primary DNA sequence for an adequate explanation of why chromosomes should band, by at least some techniques. However, the story doesn't end there; in the rest of this work we discuss how the abundance of genes in R- and T-bands, and the paucity of them in G-bands, could be directly linked with other facets of chromosome behavior such as chromatin condensation, replication timing, base composition, and recombination. If metaphase chromosome structure reflects the activities of interphase chromatin then chromosome bands can help us understand how transcribing, replicating, repairing, recombining and condensing complex genomes is coordinated. Bands also provide us with a visual framework into which the completed primary DNA sequence and the patterns inherent within it can be better understood.

Knowing where most genes are located in mammalian genomes is important for programs aimed at mapping these genomes and locating coding sequences in them. For example, it is apparent from analyses in the human genome, such as those shown in Figures 3.3 and 3.6, that sequencing of human chromosome 19 should yield a much higher density of information on protein coding sequences than a similar exercise on chromosome 18. Additionally, it has been argued that highly informative polymorphic markers chosen for scanning the genome for linkage to disorders should not be scattered evenly along chromosomes, but rather should be biased toward the most gene rich regions of the genome.[75] This can be achieved by selecting microsatellites from CpG island-rich fragments of the genome, or from GC-rich isochore fractions.[76]

REFERENCES

1. Yunis JJ and Tsai MY. Mapping of polysomal messenger RNA and heterogeneous nuclear RNA to the lightly staining G-bands of human chromosomes. Cytogenet. Cell Genet. 1978; 22:364-367.
2. Bestor TH and Coxon A. The pros and cons of DNA methylation. Curr. Biol. 1993; 3:384-386.

3. Bird AP. CpG islands as gene markers in the vertebrate nucleus. Trends Genet. 1987; 3: 342-346.
4. Larsen F, Gundersen G, Lopez R and Prydz H. CpG islands as gene markers in the human genome. Genomics 1992; 13: 1095-1107.
5. Antequera F and Bird A. Number of CpG islands and genes in human and mouse. Proc. Natl. Acad. Sci. USA 1993; 90: 11995-11999.
6. Bickmore WA and Bird AP. The use of restriction endonucleases to detect and isolate genes from mammalian cells. In: Wu R, ed. Methods in Enzymology. Vol. 216 San Diego: Academic Press, 1992:224-245.
7. Cross SH, Charlton JA, Nan X and Bird AP. Purification of CpG islands using a methylated DNA binding column. Nature Genet. 1994; 6:236-244.
8. Craig JM and Bickmore WA. The distribution of CpG islands in mammalian chromosomes. Nature Genet. 1994; 7:376-382.
9. Cross SH and Bird AP. CpG islands and genes. Curr. Opinions Genet. Develop. 1995; 5:309-314.
10. Bird A, Taggart M, Frommer M et al. A fraction of the mouse genome that is derived from islands of nonmethylated CpG-rich DNA. Cell 1985; 40:91-99.
11. Craig JM and Bickmore WA. The relationship between gene density and chromosome banding patterns in mammalian nuclei. Chromosomes Today: 1996; 12, in press.
12. Holmquist GP. Chromosome bands, their chromatin flavors and their functional features. Am. J. Hum. Genet. 1992; 51:17-37.
13. Ichikawa H, Hosada F, Arai Y et al. A NotI restriction map of the entire long arm of human chromosome 21. Nature Genet. 1993; 4:361-366.
14. Cross SH, Lee M, Clark VH et al. The chromosomal distribution of CpG islands in the mouse: evidence for genome scrambling in the rodent lineage. 1996 submitted.
15. Somssich I, Hameister H and Winking H. The pattern of early replicating bands in the chromosomes of the mouse. Cytogenet. Cell Genet. 1981; 30:222-231.
16. McQueen HA, Fantes J, Cross SH et al. CpG islands of chicken are concentrated on microchromosomes. Nature Genet. 1996; 12:321-324.
17. Auer H, Mayr B, Lambrau M and Schleger W. An extended chicken karyotype, including the NOR chromosome. Cytogenet. Cell Genet. 1987; 45: 218-221.
18. Schmid M, Enderle E, Schindler D and Schempp W. Chromosome banding and DNA replication patterns in bird karyotypes.

Cytogenet. Cell Genet. 1989; 52:139-146.
19. Antequera F and Bird AP. Unmethylated CpG islands associated with genes in higher plant DNA. EMBO J. 1988; 7:2295-2299.
20. Moore G, Abbo S, Cheung W et al. Key features of cereal genome organization as revealed by the use of cytosine methylation-sensitive restriction endonucleases. Genomics 1993; 15:472-482.
21. Polymeropoulos MH, Xiao H, Sikela JM et al. Chromosomal distribution of 320 genes from a brain cDNA library. Nature Genet. 1993; 4:381-386.
22. Rouquier S, Trask BJ, Taviaux S et al. Direct selection of cDNAs using whole chromosomes. Nucl. Acids Res. 1995; 23:4415-4420.
23. Chen-Liu LW, Huang BC, Scalzi JM et al. Selection of hybrids by affinity capture (SHAC): a method for the generation of cDNAs enriched in sequences from a specific chromosome region. Genomics 1995; 30:388-392.
24. Bond DJ and Chandley AC. Aneuploidy. Oxford monographs on Medical Genetics No. 11. Oxford University Press 1983.
25. Lyon MF, Rastan S and Brown SDM. Genetic variants and strains of the laboratory mouse. 3rd Ed. Oxford University Press 1996.
26. Gill KS, Gill BS and Endo TR. A chromosome region-specific mapping strategy reveals gene-rich telomeric ends in wheat. Chromosoma 1993; 102:374-381.
27. Elliot RW and Pazik J. An interstitial telomere array near Hba on mouse Chr 11 is a candidate for the homolog of the telomere at human 16p. Genomics 1995; 27:217-218.
28. Bernardi G. The isochore organization of the human genome. Annu. Rev. Genet. 1989; 23:637-661.
29. Mouchiroud D, D'Onofrio G, Aïssani B et al. The distribution of genes in the human genome. Gene 1991; 100:181-187.
30. Saccone S, De Sario A, Wiegant J et al. Correlations between isochores and chromosomal bands in the human genome. Proc. Natl. Acad. Sci. USA 1993; 90:11929-11933.
31. Carels N, Barakat A and Bernardi G. The gene distribution of the maize genome. Proc. Natl. Acad. Sci. USA 1995; 92:11057-11060.
32. Gardiner K, Horisberger M, Kraus J et al. Analysis of human chromosome 21: correlation of physical and cytogenetic maps; gene and CpG island distributions. EMBO J. 1990; 9:25-34.
33. Pilia G, Little RD, Aïssani B et al. Isochores and CpG islands in YAC contigs in human Xq26.1-qter. Genomics 1993; 17:456-462.
34. Fukagawa T, Sugaya K, Matsumoto K-I et al. A boundary of long-range G+C% mosaic domains in the human MHC locus: Pseudo-autosomal boundary-like sequence exists near the boundary. Genomics 1995; 25:184-191.
35. Senger G, Ragoussis J, Trowsdale J and Sheer D. Fine mapping of

the human MHC Class II region within chromosome band 6p21 and evaluation of probe ordering using interphase fluorescence in situ hybridization. Cytogenet. Cell Genet. 1993; 64:49-53.
36. Campbell RD. The human major histocompatibility complex: a 4000 kb segment of the human genome replete with genes. In: Davies KE and Tilghman SM, eds. Genome Analysis. Vol 5. Regional physical mapping. Cold Spring Habor Laboratory Press 1993:1-33.
37. Charlesworth B. Patterns in the genome. Curr. Biol. 1994; 4:182-184.
38. DeSario A, Gengl EM, Palmieri G et al. A compositional map of human chromosome band Xq28. Proc. Natl. Acad. Sci. USA 1996; 93:1298-1302.
39. Comings DE and Drets ME. Mechanisms of chromosome banding IX. Are variations in DNA base composition adequate to account for quinacrine, Hoechst 33258 and daunomycin banding? Chromosoma 1976; 56:199-211.
40. Holmquist G, Gray M, Porter T and Jordan J. Characterization of Giemsa dark- and light-band DNA. Cell 1982; 31:121-129.
41. Saitoh Y and Laemmli UK. Metaphase chromosome structure: bands arise from a differential folding path of the highly AT-rich scaffold. Cell 1994; 76:609-622.
42. Sumner AT. Electron microscopy of the parameres formed by the centromeric heterochromatin of human chromosome 9 at pachytene. Chromosoma 1986; 94:199-204.
43. Moyzis RK, Albright KL, Bartholdi MF et al. Human chromosome-specific repetitive DNA sequences: novel markers for genetic analysis. Chromosoma 1987; 95:375-386.
44. Sunkel CE and Coelho PA. The elusive centromere: sequence divergence and functional conservation. Curr. Opinions Genet. Develop. 1995; 5:756-767.
45. Royle NJ, Clarkson RE, Wong Z and Jeffreys AJ. Clustering of hypervariable mini-satellites in the proterminal regions of human autosomes. Genomics 1988; 3:352-360.
46. Singer MF. Highly repeated sequences in mammalian genomes. Int. Rev. Cytol. 1982; 76:67-113.
47. Sun L, Paulson KE, Schmid CM et al. Non-Alu family interspersed repeats in human DNA and their transcriptional activity. Nucl. Acids Res. 1984; 12:2669-2690.
48. Hwu HR, Roberts JW, Davidson EH and Britten RJ. Insertion and/or deletion of many repeated DNA sequences in human and higher ape evolution. Proc. Natl. Acad. Sci. USA 1986; 83: 3875-3879.
49. Manuelidis L and Ward DC. Chromosomal and nuclear distribu-

tion of the HinDIII 1.9 kb human DNA repeat segment. Chromosoma 1984; 91:28-38.
50. Korenberg JR and Rykowski MC. Human genome organization. Alu, Lines, and the molecular structure of metaphase chromosome bands. Cell 1988; 53:391-400.
51. Boyle AL, Ballard SG and Ward DC. Differential distribution of long and short interspersed element sequences in the mouse genome: chromosome karyotyping by fluorescence in situ hybridization. Proc. Natl. Acad. Sci. USA 1990; 87:7757-7761.
52. Soriano P, Meunier-Rotival M and Bernardi G. The distribution of interspersed repeats is non-uniform and conserved in the mouse and human genomes. Proc. Natl. Acad. Sci. USA 1983; 80: 1816-1820.
53. Chen TL and Manuelidis L. SINEs and LINEs cluster in distinct DNA fragments of Giemsa band size. Chromosoma 1989; 98:309-316.
54. Thomsen PD and Miller JR. Pig genome analysis: differential distribution of SINE and LINE sequences is less pronounced than in the human and mouse genomes. Mamm. Genome 1996; 7:42-46.
55. Chen EY, Liao Y-C, Smith DH et al. The human growth hormone locus: Nucleotide sequence, biology, and evolution. Genomics 1989; 4:479-497.
56. Edwards A, Voss H, Rice P et al. Automated DNA sequencing of the human HPRT locus. Genomics 1990; 6:593-608.
57. Porta G, Zucchi I, Hillier L et al. Alu and L1 sequence distributions in Xq24-q28 and their comparative utility in YAC contig assembly and verification. Genomics 1993; 16:417-425.
58. Yoshiura K-I, Kubota T, Soejima H et al. A comparison of GC content and the proportion of Alu/Kpn-repetitive sequences in a single dark- and light-band region from a human chromosome. Genomics 1994; 20:243-248.
59. Natsoulis G, Thomas W, Roghmann M-C et al. Ty*1* transposition in *Saccharomyces cerevisiae* is nonrandom. Genet. 1989; 123:269-279.
60. Scherdin U, Rhodes K and Breindl M. Transcriptionally active genome regions are preferred sites for retrovirus integration. J. Virol. 1990; 64:907-912.
61. Capel J, Montrero LM, Martinez-Zapater and Salinas J. Non-random distribution of transposable elements in the nuclear genome of plants. Nucleic Acids Res. 1993; 21:2369-2373.
62. Mason JM and Biessmann H. The unusual telomeres of *Drosophila*. Trends Genet. 1995; 11:58-62.
63. Kazazian HH, Wong C, Youssoufian H et al. Haemophilia A resulting from de novo insertion of *L1* sequences represents a novel mechanism for mutation in man. Nature 1988; 332:164-166.

64. Wallace MR, Andersen LB, Saulino AM et al. A *de novo Alu* insertion results in neurofibromatosis type I. Nature 1991; 353:864-866.
65. Margalit H, Nadir E and Ben-Sasson SA. A complete Alu element within the coding sequence of a central gene. Cell 1994; 78:173-174.
66. Clark SJ, Harrison J and Frommer M. CpNpG methylation in mammalian cells. Nature Genet. 1995; 10:20-27.
67. Miller OJ, Schnedl W, Allen J and Erlanger BF. 5-Methylcytosine localized in mammalian constitutive heterochromatin. Nature 1974; 251:636-637.
68. Lewis JD, Meehan RR, Henzel WJ et al. Purification, sequence, and cellular localization of a novel chromosome protein that binds to methylated DNA. Cell 1992; 69:905-914.
69. Schmid M, Haaf T and Grunert K. 5-azacytidine-induced undercondensations in human chromosomes. Hum. Genet. 1984; 67:257-263.
70. Vongs A, Kakutani T, Martienssen RA and Richards EJ. Arabidopsis thaliana DNA methylation mutants. Science 1993; 260:1926-1928.
71. Nasir J, Maconochie MK and Brown SDM. Co-amplification of L1 line elements with localized low copy repeats in Giemsa dark bands: implications for genome organization. Nucl. Acids Res. 1991; 19:3255-3260.
72. Koop BF and Nadeau JH. Pufferfish and a new paradigm for comparative genome analysis. Proc. Natl. Acad. Sci. USA 1996; 93:1363-1365.
73. Bird AP. Gene number, noise reduction and biological complexity. Trends Genet. 1995; 11:94-100.
74. Duret L, Mouchiroud D and Gautier C. Statistical analysis of vertebrate sequences reveals that long genes are scarce in GC-rich isochores. J. Mol. Evol. 1995; 40:308-317.
75. Antonarakis SE. Genome linkage scanning: systematic or intelligent? Nature Genet. 1994; 8:211-212.
76. Gyapay G, Morissette J, Vignal A et al. The 1993-1994 Genethon human genetic linkage map. Nature Genet. 1994; 7:246-339.
77. Ikemura T, Wada K-N and Aota S-I. Giant G+C% mosaic structures of the human genome found by arrangement of GenBank human DNA sequences according to genetic positions. Genomics 1990; 8:207-216.
78. Collins F and Weissman S. The molecular genetics of human hemoglobins. In: Cohn WE and Modave K, eds. Progress in Nucleic Acids Research and Molecular Biology, Vol. 31. Orlando: Academic Press, 1984:315-462.
79. Toguchida J, McGee TL, Paterson JC et al. Complete genomic sequence of the human retinoblastoma susceptibility gene. Genomics 1993; 17:535-543.

80. Chen EY, Zollo M, Mazzarella R et al. Long-range sequence analysis in Xq28: thirteen known and six candidate genes in 219.4 kb of high GC DNA between the RCP/GCP and G6PD loci. Hum. Mol. Genet. 1996; 5:659-668.
81. Martin-Gallardo A, McCombie WR, Gocayne JD et al. Automated DNA sequencing and analysis of 106 kilobases from human chromosome 19q13.3. Nature Genet. 1992; 1:34-39.
82. Iris FJM, Bougueleret L, Prieur S et al. Dense Alu clustering and a potential new member of the NFκB family within a 90 kilobase HLA class III segment. Nature Genet. 1993; 3:137-145.
83. Nicholls RD, Fischel-Ghodsian N and Higgs DR. Recombination of the human alpha-globin gene clusters. Sequence features and topological constraints. Cell 1987; 49:369-378.
84. Fischel-Ghodsian N, Nicholls RD and Higgs DR. Unusual features of CpG-rich (HTF) islands in the human α globin complex: association with nonfunctional pseudogenes and presence within the 3' portion of the ζ gene. Nucl. Acids Res. 1987; 15:9215-9225.
85. Morton NE. Parameters of the human genome. Proc. Natl. Acad. Sci. USA 1991; 88:7474-7476.

CHAPTER 4

The Chromatin Structure of Chromosome Bands

The primary DNA sequence is wrapped into the form of the chromosome by a series of hierarchical interactions with chromosomal proteins. Several of the techniques of chromosome banding suggest that there are both qualitative and quantitative variations in the nature of the interaction of DNA with proteins along the length of metaphase chromosomes. For example, chromosome bands can be generated by nuclease digestion of chromosomes and by G-banding protocols that involve a proteolysis step. There is also evidence that sulphydryl and disulphide groups in proteins may be connected with the mechanism of G-banding. Banding has also been achieved with other agents that affect protein structure such as urea, guanidine hydrochloride, potassium permanganate, ammonium sulphate, and fluorescamine.[1]

Since, as discussed in the previous chapter, there are fundamental differences in the characteristics of the primary DNA sequence along the chromosome length, it is not surprising to find that there are also differences in the chromosomal proteins found within chromosome bands of different flavors. Indeed, examining such protein distributions can give us clues as to possible functions of the protein in question. For example, a protein involved in the formation of an inactive-heterochromatin-like state would be unlikely to turn up mainly in T-band regions of chromosomes assuming that metaphase chromatin structure is related to that at interphase in at least some respects. Below we discuss chromosomal proteins whose distributions along the chromosome length relate to chromosome bands and examine how their proposed function(s) can be related to their distribution.

CORE HISTONES

The first and best understood interaction of DNA and proteins in chromosomes is that which occurs with the histones to form nucleosomes (Fig. 1.1). The individual histone molecules of the core nucleosome can be biochemically modified in a variety of ways including methylation, phosphorylation, acetylation, ubiquitination and ADP-ribosylation. The acetylation of lysine residues in the N-terminal domains of histones H3 and H4 is the modification that has attracted the most intensive interest and examination. The N-terminal tails of the core histones, protruding on the outside of the nucleosome, are the parts of the histone molecules most likely to readily contact other proteins such as transcription factors, or even nucleosome destabilizing or disrupting protein complexes. The consequence of acetylation may be to facilitate access of proteins such as transcription factors to the DNA by decreasing the ability of the histones to constrain the path of DNA and by destabilizing internucleosomal contacts.[2,3] One manifestation of this is the increase in DNase I sensitivity of chromatin after the induction of histone acetylation through the use of deacetylase inhibitors.[4]

Newly synthesized histones are usually N-terminally acetylated at two lysine residues by cytoplasmic histone acetyltransferases and are transported to the nucleus and assembled onto newly replicated DNA in this form.[5,6] After assembly, histones can then be deacetylated or further acetylated throughout the rest of the cell cycle by nuclear acetyltransferase and deacetylase enzymes.[7] Acetylation at the different lysine residues of H3 and H4 tends to occur in the same general, but not strictly observed, order in many species. With the exception of *Drosophila*,[8] the order in which acetylation of H4 occurs tends to be firstly lysine 16 followed by 8 or 12, with lysine 5 being acetylated principally only in the most highly modified (tri- and tetra-acetylated) forms. What is the controlling factor in this preferred order of use? We are just beginning a new and exciting phase in our understanding of histone acetylation with the identification and isolation of one nuclear histone acetyltransferase and one deacetylase.[9,10] There is some indication from the initial studies of the activity of these enzymes that in vivo they may have some specificity for particular lysine residues of either H3 or H4. Hence it is likely that there will be several other as yet unidentified enzymes with different or overlapping specificities—histone acetyltransferase A is not an essential gene, indicative of functional redundancy. Additionally or alternatively, the activity of these enzymes may be modulated by their chromosomal location and the proteins with which they interact.[9]

Immunofluorescence of mammalian metaphase chromosomes with antibodies raised against each of the acetylated forms of H4 has shown that the differently modified forms are found preferentially in different regions of the chromosome suggesting that each may have a different function. Mono-acetylated (lys 16) H4 is found throughout euchromatin whereas acetylation at lysines 8 and 12 occurs mainly in R-bands. Acetylation at lysine 5 is found, in the main, only in the most highly modified (tri- and tetra-acetylated) forms that account for 4% of histone H4. Antibodies to this H4 isoform produce a good banding pattern on metaphase chromosomes, especially from cells that have been briefly exposed to histone deacetylase inhibitors. Bright immunofluorescence is seen over R-band regions of the karyotype and only faint staining is seen to G-bands.[11] Hence the distribution of histone acetylation on mammalian metaphase chromosomes mirrors that of genes, and the same may be true in plants.[12] The assumption in these analyses is that the pattern of acetylation revealed at metaphase by immunofluorescence reflects at least some aspects of the interphase chromatin state. However, it is known that levels of H4 acetylation (especially of the most highly modified forms) decrease as cells enter mitosis.[13,14]

Is the concentration of nucleosomes containing the most highly acetylated forms of H4 into R-bands just a passive marker for the concentration of genes and hence transcription in these regions? Or, is the acetylation of histones in these regions necessary for the transcription of these genes and hence, active contribution to the chromatin environment of different bands? Consistent with immunofluorescence studies, interphase chromatin fractions from active CpG islands have highly acetylated core histones.[15]

Immunoprecipitation studies also show that acetylation is limited not only to the promoter or coding regions of genes but is spread over entire chromatin domains that contain genes capable of being transcribed in the cells being examined.[16] However, immunofluorescence of insect polytene (interphase) chromosomes has shown that sites of very high levels of transcription (puffs) are not associated with especially high levels of H4 acetylation[17] and immunoprecipitation experiments in mammalian cells has indicated that the levels of H4 acetylation are only slightly higher along genes (both active and silent) than in bulk chromatin, but are a lot higher than heterochromatin.[18] More importantly, they detected no changes in steady state H4 acetylation accompanying the switching on or off of genes. This suggests that although acetylation of H4 is one of the features that make chromatin permissive for transcription, it plays little direct role in the day to day control of transcription.

The same may not be true of acetylation of histone H3. Genetic analyses in the yeast *S. cerevisiae* suggest that acetylation of histones H3 and H4 may have different consequences. Whereas N-terminal deletions of H3 result in hyperactivity of some genes[19] similar mutations in histone H4 decrease the activation of these genes. However, the main phenotype is a loss of genetic silencing at several sites in the genome (for example at the silent mating type loci). Swapping the N-terminus of H3 onto H4 results in a more severe effect on silencing than a deletion of the N-terminus of H4.[6] One interpretation of these results is that H3 acetylation may have more of a direct role in inducible transcription from euchromatin whereas histone H4 acetylation may block the formation of heterochromatin (chapter 6).

Thus far we have considered mainly the steady state levels of histone acetylation at interphase or metaphase, often after inhibition of deacetylation. However, histone acetylation/deacetylation is very dynamic and there are different histone populations in which the acetate groups turn over with different half lives. A minority (<15%) of histones turn over their acetate groups rapidly with half-lives of less than ten minutes and this population appears to include the most highly acetylated histone forms.[20] Consistent with a concentration of highly acetylated H4 in the chromatin in the proximity of genes, the acetate groups on H4 molecules associated with genes (whether transcriptionally on or off) turn over faster than those associated with bulk chromatin.[18] Do these differences in the dynamics of histone modification reflect the ease of access for acetyltranferase and deacetylase enzymes in transcriptionally competent chromatin or is it a reflection of the spatial distribution of the enzymes in the nucleus? The answer to this question appears to be the latter. Both the recently cloned nuclear acetyltranferase and the deacetylase were found to be homologous to, or related to, known transcriptional regulators, providing a ready means to explain the targeting of histone acetylation to particular regions of the genome.[9,10] Histone acetyltransferase A is a homologue of a transcriptional adapter protein Gcn5p involved in activation of certain genes and, aside from its catalytic domain, it contains a domain (the bromodomain) that is found in other chromatin-associated proteins that have a role in transcriptional activation.[9] The deacetylase is related to a transcriptional regulator Rpd3 isolated in a screen for transcriptional repressors. Linking the enzymatic machinery for reversible histone acetylation to components of transcriptional regulators answers some of the questions of how histone acetylation is targeted to the immediate vicinity of genes—especially their 5' ends. However, it is hard to imagine how these same enzymes are capable of acetylating or deacetylating whole chromosome domains including inter-genic nucleosomes.[11,16,18] In addition, histones within the chromatin of metaphase chromosomes seem to be capable of being deacetylated, indicating that deacetylase enzyme activity persists in metaphase cells[14]—a stage of the cell cycle when most transcription factors appear to be excluded from the chromosome (chapter 5). The isolation of other enzymes with similar activities but different localizations may explain these discrepancies.

We know much less about other forms of histone modification. Lysines in the C-terminal tails of histones H2A and H2B can be reversibly ubiquitinated. There is some suggestion that nucleosomes from active regions of the genome are enriched in ubiquitinated H2A and particularly H2B.[21] Studies of metaphase chromosomes, however, provide no further insight into this since the ubiquitin moieties are removed as cells progress through prophase and do not reappear until anaphase.[22] ADP-ribosylation of histones preferentially occurs on acetylated histones and so might be expected to occur more frequently in R-bands.[23]

Regulatable phosphorylation of serines 10 and 28 in the N-terminal tails of histones occurs but the role of this is not understood.[24] A role in chromosome condensation might be suggested by the observation that phosphorylation of H3 peaks at metaphase.

There are a variety of histone-like molecules associated with chromosomes. One specialized histone H3 variant with a very restricted location on metaphase chromosomes is CenP-A that is found in the nucleosomes of active centromeres (chapter 6).[25]

LINKER HISTONES

Histone H1 facilitates the packaging of nucleosomes into a 30 nm fiber and some, but not all, studies indicate that it is present in lower levels in active as compared to inactive chromatin. This has led to the suggestion that is has a generally repressive effect on transcription.[26] Consistent with this, H1 is depleted in chromatin at CpG islands, at least for those islands associated with active genes where the core histones are hyperacetylated.[15] Conversely, H1 is enriched in nucleosomes containing ^{5Me}C.[27] Histone hyperacetylation decreases the amount of H1 in nucleosomal arrays[28] thus, as might be expected, one immunofluorescence study has indicated that several H1 subtypes are more abundant in G-bands than R-bands at metaphase and that changing the acetylation of core histones enhanced this banding profile.[29] However, other studies have been unable to find differences in H1 abundance between different bands and this dichotomy remains unresolved.[30]

Like the core histones, linker histones are also subject to post-translational modification and, like H3, phosphorylation of H1 reaches a peak at metaphase, and declines rapidly at anaphase. H1 phosphorylation is catalyzed by a cdc2/cyclin kinase,[31] and the dephosphorylation, by a phosphatase that is a chromosomal component and that may be protein phosphatase 1.[32] It is unclear how the H1 phosphorylation state relates to the process of mitotic chromosome condensation since intuitively phosphorylation (addition of negatively charged groups) might be thought to weaken the interaction of H1 with DNA and hence result in decondensation. Indeed, dephosphorylation of histones H1 and H3 has been shown to occur in vitro during the preparation of metaphase chromosomes with no resulting decondensation.[33] In *Tetrahymena,* H1 phosphorylation is not correlated solely with mitosis but rather correlates best with chromatin decondensation.[34] One proposition is that phosphorylation of H1 during mitosis indeed weakens H1-DNA interactions, but that by doing so it allows higher order chromatin interactions of H1 to occur[35]—as we shall see, H1 seems to interact with the chromosome scaffold. In vitro studies also suggest that linker histones are not needed at all for mitotic chromosome formation.[36] The most conclusive evidence for this, however, comes from a genetic knockout of the micronuclear linker histone gene (MLH) in *Tetrahymena.*[37] Cells lacking MLH have an apparently normal mitosis albeit with less condensed chromosomes than normal.

H1 is not the only form of linker histone and the choice of linker can be modulated during development. Not all of these alternative molecules seem to be capable of being phosphorylated by cdc2 kinase.[4]

HIGH MOBILITY GROUP NONHISTONE PROTEINS

High mobility group (HMG) proteins are communally named on biochemical grounds because they were identified as the abundant, small, charged nonhistone components of chromatin. We now know that they fall into at least three distinct families that are largely unrelated to each other in terms of protein sequence, but still the common prefix HMG has stuck. The first family is characterized by HMG1 and 2, whose DNA binding domains gave rise to the canonical term 'HMG box' for a motif found in proteins as diverse as UBF, ABF2, and SRY. We know little about the distribution of HMG1/2 on chromosomes or their functions, though this family of proteins are often thought of as architectural factors that influence chromatin structure through the formation or stabilization of DNA bends.[38] Some mitotic chromosome banding has been reported with HMG2,[39] however, no differences in the abundance of HMG1/2 in transcribed versus untranscribed chromatin could be detected by immunoprecipitation from chicken chromosomes.[26] A homologue of HMG1/2 in *Drosophila* (HMG-D) is associated with mitotic chromosomes during embryogenesis (from which H1 is absent).[40]

HMG14/17 are members of a second family of 'HMG' proteins with no homology to HMG1/2. They are preferentially associated with the nucleosomes from active chromatin but their distribution on metaphase chromosomes is not known.[26]

HMGI/Y (the products of alternate splicing of the same gene) represent the last class of HMG proteins, which like HMG1/2 type proteins are thought to bend DNA.[41] They do not have an HMG box like HMG1/2, but instead have a DNA binding domain termed an 'AT hook.' HMGI/Y are a rare example of proteins whose distribution on metaphase chromosomes, as judged by immunofluorescence, is hard to reconcile with their proposed function. Antibodies directed against these proteins produce clear G-banding patterns on mammalian metaphase chromosomes.[42] This is consistent with the ability of these proteins to bind to AT-rich DNA, and the production of G-bands on mammalian chromosomes by AT-specific fluorochromes. These lines of evidence suggest that these proteins serve some architectural role in the folding of certain chromosomal domains. However, HMGI/Y, although not a transcriptional activator itself, has been shown to have a role in gene expression—they facilitate the binding of transcription factors such as NF-κB and ATF2 in the regulation of interferon (IFN) β gene expression.[43] If HMGI/Y also had a role in the transcription of a large number of other genes, one might expect to see it concentrated in the gene rich R-band regions of the genome (at least in interphase)—the human IFNB1 and 3 genes are located in R-bands. One way to reconcile these two opposing views is to propose that this protein fulfills a general role in the packing of chromatin in the metaphase chromosome. Additionally their ability to bend DNA has been utilized by a few genes in the control of their gene expression to bring about the spatial juxtaposition of sequence-specific transcription factor binding sites.

PROTEINS OF THE MITOTIC CHROMOSOME SCAFFOLD

The radial loop/scaffold model of chromosome structure (Fig. 4.1) proposes that higher order chromosome packaging arises through the arrangement of the 30 nm chromatin fiber into loops tethered to a proteinaceous structure that may be either rigid, discontinuous or dispersed. The term chromosome scaffold was coined to describe such a residual structure, seen by electron microscopy (EM) of mammalian mitotic chromosomes that had had most proteins biochemically extracted from them.[44] This morphologically defined chromosome scaffold runs centrally down the length of the chromosome and retains the approximate size and shape of the unextracted chromosome with clearly visible paired sister chromatids, often still attached to each other at the position of the centromere (Fig. 1.3). This scaffold is surrounded by a halo that EM shows is comprised of loops of DNA emanating from the

A

B

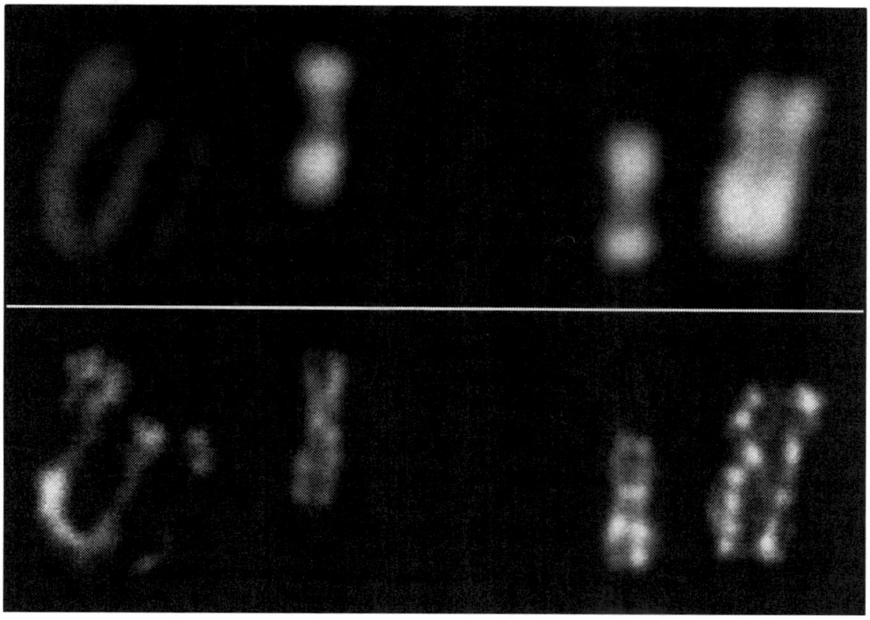

Fig. 4.1. Topoisomerase II decorates the core of metaphase chromosomes. Four human metaphase chromosomes prepared by cytocentrifugation and incubated with an antibody detecting topoisomerase IIα prior to fixation (courtesy of Ian Hickson, Oxford). DAPI staining is shown in panel A and the pattern of immunofluorescence in B. Note that the pattern of immunofluorescence is not uniform along the chromosome length and that it is symmetrical on the two sister chromatids.

scaffold. The existence of such a central organizing center in mitotic chromosomes is strengthened by the observations of similar axial elements in swollen but unextracted chromosomes[45] and in silver stained whole chromosomes.[46]

The mitotic scaffold contains only 3-4% of the chromosome mass and its two major protein components are Sc I (170 kD) and Sc II (135 kD).[47] It is therefore tempting to endow these proteins with structural roles. Sc I is topoisomerase II and the first evidence for a mitotic scaffold seen with the light microscope was the pattern of topoisomerase II immunofluorescence on the long axis of chromosomes swollen in hypotonic buffer (Fig. 4.1).[48] A similar localization is seen with Sc II, a member of the SMC family of proteins.[49] These two types of protein are also among the few proteins tightly associated with chromosomes that have been condensed in a cell free system.[50,51]

Topoisomerase II alters DNA topology, catalyzing the passage of one intact DNA helix through a transient double-stranded break in a second DNA helix in an ATP-dependent manner. It has been implicated in many aspects of nucleic acid metabolism that are important for the successful execution of mitosis: separation of newly-replicated DNA,[52] premitotic chromatin condensation and anaphase sister chromatid separation.[53,54]

The precise role of topoisomerase II in maintaining the structural integrity of the mitotic chromosome is called into doubt by some, but not all, immunodepletion experiments in *Xenopus* egg extracts[50,55] and by cell cycle and immunofluorescence analysis in *Drosophila* embryos.[56] The situation is further complicated by the fact that in vertebrates there are two isoforms of topoisomerase II encoded by separate genes. It is the protein encoded by the TOP2α gene (p170) that is a major part of the mitotic chromosome scaffold. This enzyme is only found in abundance in dividing cells and most of it copurifies with the chromosomes. Similarly, most of the chromosomal fraction of p170 can be recovered in the scaffold fraction.[57,58] These analyses, together with the patterns of immunofluorescence of topoisomerase on mitotic chromosomes, are strong arguments in favor of at least a subpopulation of this enzyme being located at the chromosome scaffold and having a function there that is both enzymatic and structural. More careful analy-

Color Figures

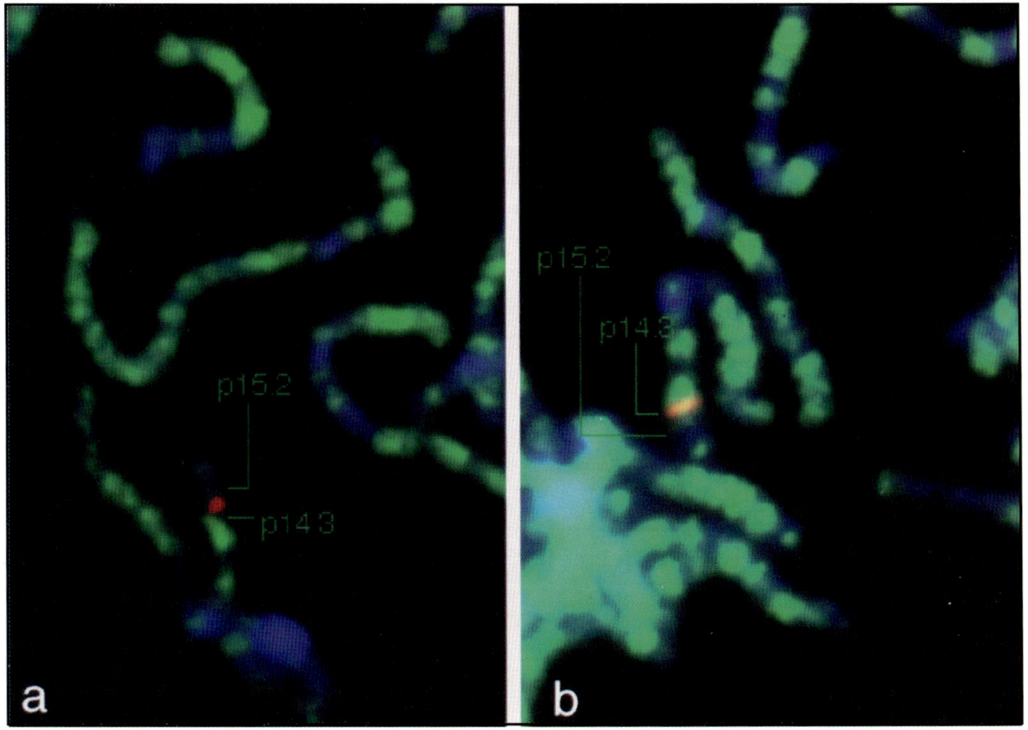

Fig. 2.3. FISH and high-resolution banding. FISH signal (red or orange) from YACs mapping to human chromosome 11 is detected on chromosomes stained with DAPI (blue) and in which BrdU has been incorporated into late replicating DNA (green). (a) The YAC maps to 11p15.1 (an early replicating R-band) and hence the hybridization signal is red since there is no BrdU present in this region of the chromosome. (b) The YAC maps to 11p14.3 (a late replicating G-band) and so the hybridization signal appears orange where red is superimposed on green (from the underlying BrdU). Reproduced courtesy of Judy Fantes, MRC Human Genetics Unit, Edinburgh.

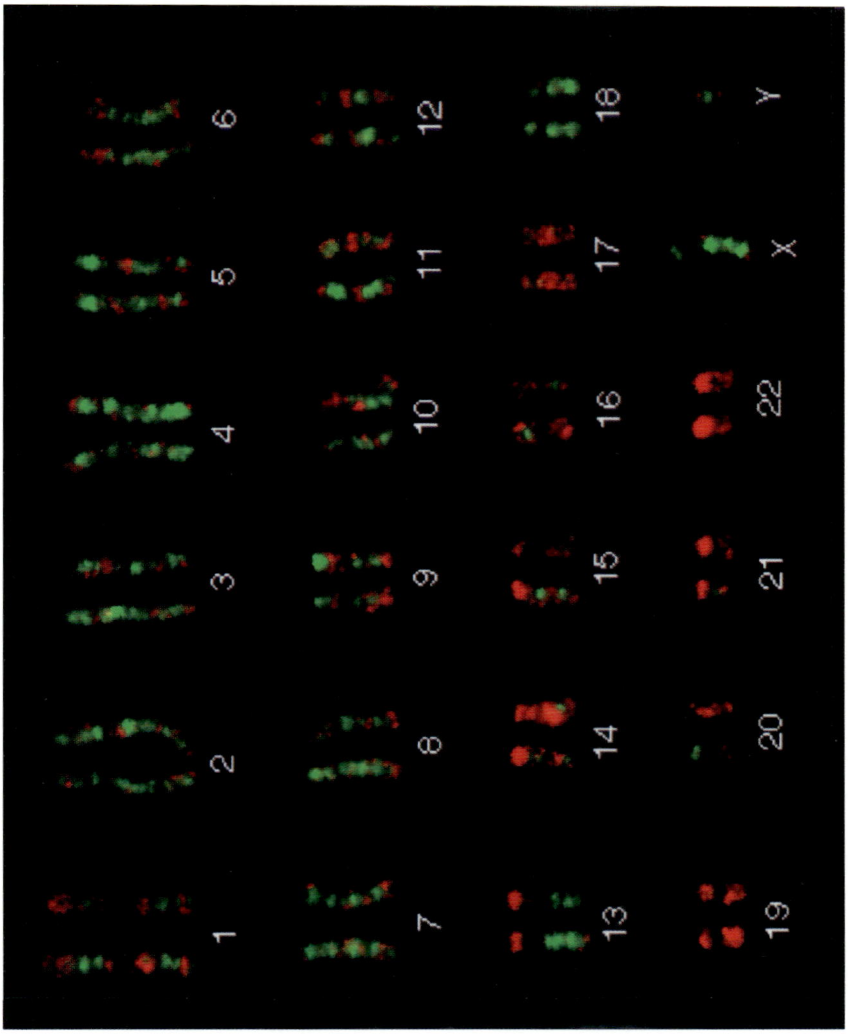

Fig. 3.3. Distribution of CpG islands in the human karyotype. CpG islands were purified as small BstU1 fragments (Figs. 3.1 and 3.2), biotin-labeled and hybridized to human metaphase chromosomes that had been replication banded with BrdU as described in reference 8. CpG islands are visualized in red. Late-replicating G-bands were detected with anti-BrdU-FITC (green).

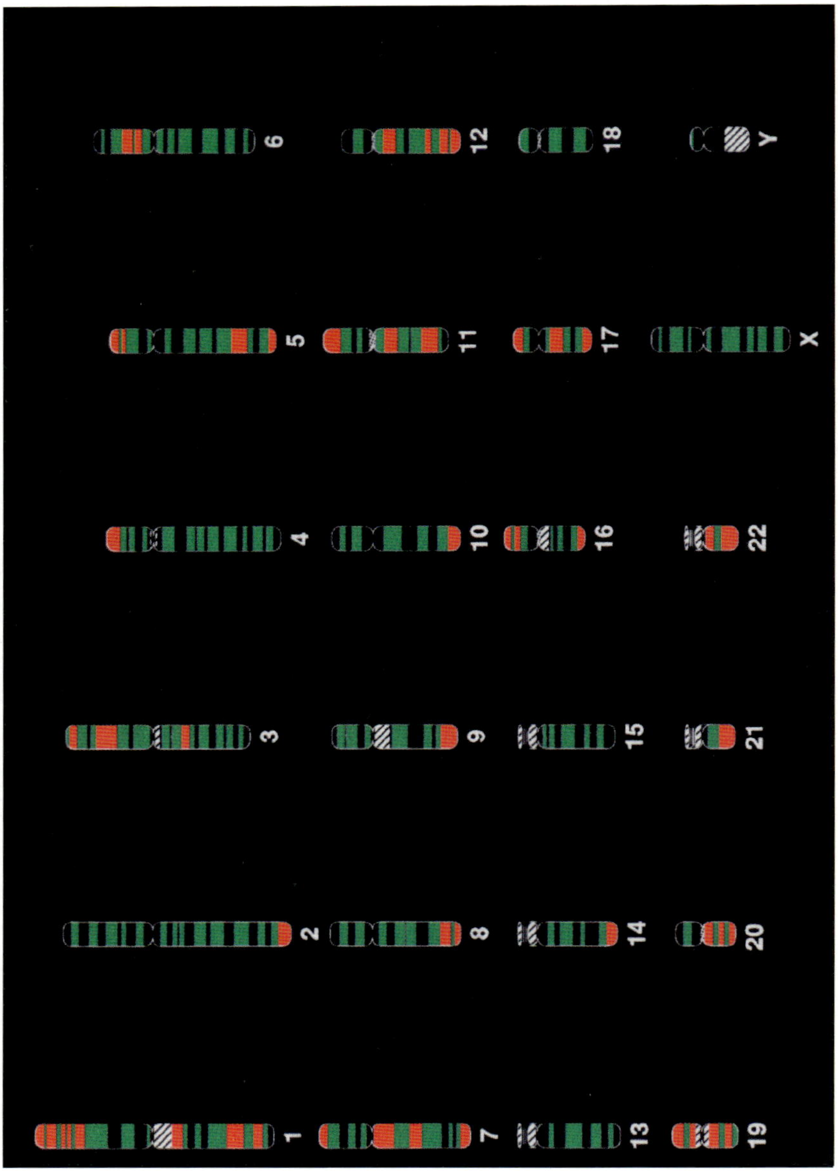

Fig. 3.4. Human chromosome ideogram. G-bands are shown in green, T-bands in red and other R-bands in black, at a resolution of 400 bands per genome. Adapted from ref. 12.

sis of the patterns of immunofluorescence along the scaffold show that topoisomerase II is not uniformly located along the chromosome axis. It appears to be concentrated in particular regions of the scaffold (Fig. 4.1)[48] with concentrations at centromeric heterochromatin until anaphase.[59] In euchromatin there is stronger topoisomerase II immunofluorescence in G-bands than R-bands.[60] We do not yet know the biological significance of these concentrations of topoisomerase II or whether the protein serves any purely structural role aside from its enzymatic one. For example, does the apparent concentration of the protein in the G- and C-band regions of the mitotic scaffold indicate that chromosome loop anchoring sites are more frequent there than those in R-bands or are they of a different type? Since topoisomerases can be modified by kinases at mitosis it will be important to determine which phosphorylated forms of topoisomerase II are recognized by the different antibodies used to detect the presence of topoisomerase II and whether differently modified forms of the protein are found on chromosomes at different times and places.[61] For example, phosphorylation of topoisomerase II, such as occurs at mitosis, enhances its ability to form multimers, and to interact with SARs and hence, might be important in the structural organization of mitotic chromosomes.[62,63] There may also be a concentration of topoisomerase I at centromeres of mammalian mitotic chromosomes prior to anaphase.[64]

The other predominant protein of the mitotic scaffold ScII has been shown to be a member of the SMC (structural maintenance of chromosomes) protein family. The structure of these proteins includes NTP binding and coiled-coil domains reminiscent of mechanochemical motor proteins which, in the case of ScII, is satisfying if one envisages that the scaffold plays a role in the dynamic process of condensing chromatin for mitosis. A role for ScII in establishing and maintaining mitotic chromosome condensation is clearly indicated genetically by the phenotypes of SMC family member mutants in yeasts[65-67] and biochemically by immunodepletion from *Xenopus* egg extracts.[51] The pattern of immunofluorescence with anti-ScII antibodies on vertebrate mitotic chromosomes is very similar to that of topoisomerase II—the chromosome axes are lit up and centromeres are particularly brightly stained. However, whereas topoisomerase II is a component of the interphase matrix as well as the mitotic scaffold, ScII is only found on mitotic scaffolds.[49] Clearly we need to find out how ScII is recruited to the mitotic scaffold and whether, like the *Xenopus* proteins XCAPC and E, it forms heterodimers with other SMC protein family members.

The biology of the two most abundant proteins of the mitotic chromosome scaffold indicate that the scaffold is important in forming and maintaining the condensed structure of the chromosome which may be needed to protect chromatin from damage during the dynamics of mitosis. Additionally, the scaffold is important for functional aspects of mitotic chromosome biology; the need to decatenate replicated sister chromatids—for which the process of condensation may play a direct role, and also the need to maintain sister chromatid cohesion until the metaphase to anaphase transition. Genetic interactions in yeasts and biochemical copurification experiments indicate that topoisomerase II and ScII probably cooperate in at least some of these roles[68,69] though these findings seem a little at odds with the loose association of ScII with the interphase nuclear matrix. Indeed, as we will discuss, the whole field of the relationship between the interphase nuclear matrix/scaffold/skeleton and the mitotic scaffold is fraught with controversy.

DNA SEQUENCES ASSOCIATED WITH THE MITOTIC SCAFFOLD

Are specific sites along the DNA (at the bases of loops in the radial loop/scaffold model) associated with the chromosome scaffold and if so, what are they? Such sequences are generally referred to as SARs (scaffold attachment regions) and in the main they have been examined by analyzing DNA remaining bound to insoluble proteins after biochemical extraction procedures, more often from nuclei rather than from mitotic chromosomes.[70,71] Formally, no individual SAR sequence has been shown to associate with the visible mitotic scaffold, however, examination of the bulk characteristics of SARs has been argued to support this conclusion. SARs generally contain several oligo(dA)-oligo (dT) tracts that have been shown to be responsible for cooperative interactions with topoisomerase II in vitro.[72] In vitro SARs fluoresce brightly with daunomycin (reflecting their AT-richness) and it has been argued that this dye can therefore be used to optically trace the path of aligned SARs (the AT queue) along the scaffold at the core of metaphase chromosomes.[60] Using the ability of the nonfluorescent dye methyl green, which also preferentially binds AT-rich DNA, to quench the fluorescence from the general fluorochrome, YOYO-1 loop DNA emanating from the central AT queue was delineated. Combining these staining procedures striking reciprocal banding patterns on mammalian metaphase chromosomes were obtained—daunomycin highlighting G-bands and YOYO/methyl green R-bands. The daunomycin staining also colocalized with the pattern of topoisomerase II immunofluorescence both along the length and across the width of the chromosomes.[60]

How might these banding patterns be interpreted? It might simply be that G-bands have smaller loops, and hence more frequent scaffold attachments than R-bands (Fig. 4.2). However, on more careful examination it was concluded that these staining patterns actually resulted from differences in the path of the SARs between different types of band with the chromosome scaffold being more tightly coiled within G-bands and straighter within R-bands. Hence there would be a more lateral arrangement of the AT queue in G-bands and a more longitudinal one in R-bands. This may be compounded by loops actually being longer in R-bands per se, though not necessarily so. If this is the case it provides a nice way of explaining why R-bands appear to become engulfed in G-bands as chromosomes become progressively more condensed (chapter 5).

A coiling of the chromosome scaffold is also consistent with the helicity that can be seen in silver stained chromosome cores[73] and with the radial loop/scaffold models in which aggregates of loops form rosettes that coil along a helical path (chapter 1). Careful optical sectioning of patterns of topoisomerase II immunofluorescence along the mitotic scaffold also reveals a helicity that may have opposite handedness between sister chromatids.[74]

SARs have numerous sites capable of binding to two proteins that we have met before—histone H1[75] and HMGI/Y. The latter is a protein whose immunofluorescence pattern follows the path of the AT queue, explaining its apparent visible enrichment in

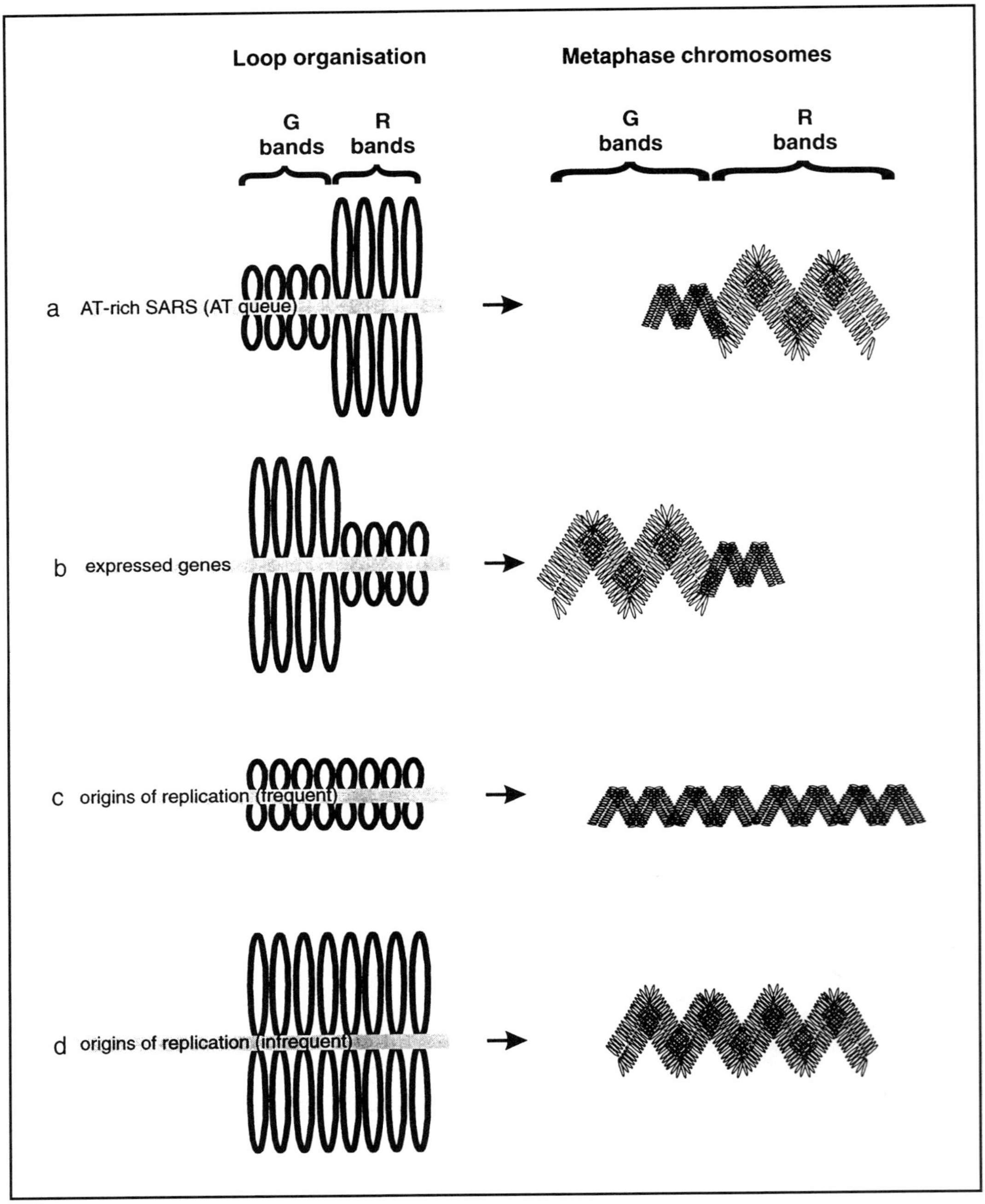
Loop organisation
Metaphase chromosomes
G bands
R bands
G bands
R bands
a AT-rich SARS (AT queue)
b expressed genes
c origins of replication (frequent)
d origins of replication (infrequent)

Fig. 4.2.

G-bands.[60] HMGI/Y is enriched in H1-depleted chromatin and it can displace H1 from SARs, relieving transcriptional repression of linked genes—hence completing the link between the localization of this protein along metaphase chromosomes and its role in transcription.[76] Experiments that use an artificial protein containing multimers of the AT hook domain from HMGI/Y have been designed to illustrated the role of SARs as cis acting elements in the determination and maintenance of metaphase chromosome shape.[77] Titration of SARs by such a protein results in the in vitro assembly of misshapen spherical condensed balls of chromosomes.

It is an appealing idea that the topological constraint of DNA into loops anchored to some type of chromosome scaffold should naturally organize the genome of an organism into functional units. Despite the satisfying simplicity of this idea, lines of evidence in support of it have been largely circumstantial and controversial. Of the various proposals are those that suggest that a chromosome scaffold/matrix might be associated with important functional elements of chromosomes such as centromeres, replication origins, or sites of transcription[71,78,79] and even with other chromatin-related activities such as histone acetylation and deacetylation.[80]

Alternatively, a chromosome scaffold, apparent within chromosomes at mitosis, may serve no purpose other than to aid in the spatial management of the chromosomal DNA molecule and may

Fig. 4.2 (opposite). Chromosome loops and chromosome bands. The role that the size of loops anchored to a mitotic chromosome scaffold might have in banding patterns is indicated dependent on the details of a radial loop/scaffold model of chromosome structure envisioned.

(a) SARs are anchored to the chromosome scaffold and they may be more frequent in G- rather than R-bands. This might translate into a thinner metaphase chromatid fiber in G-bands if the path of the scaffold is the same in G- and R-bands—which may not in fact be the case.[60]

(b) The aggregation of transcription complexes at the bases of loops directs the condensation of mitotic chromosomes,[79] thus G-bands should have much larger loops than R-bands and T-bands even smaller loops. The metaphase chromatid fiber would be thinner in R-bands than in G-bands and a gene-rich chromosome such as HSA19 would be thinner than a gene poor chromosome, e.g., HSA18.

(c and d) Origins of replication colocalize with attachments to the mitotic scaffold and are uniform along the length of euchromatin and most heterochromatin for any particular chromosome.[81,83,85] In some organisms early embryonic stages are characterized by very quick cell cycles and this is accompanied by frequent initiation of replication and consequently longer, thinner metaphase chromosomes (c) than chromosomes from cells at later stages of development in which replicon size is larger (d).[83,84]

be completely independent of other chromosomal activities. We will consider these propositions in light of the predictions that can be made, within each model, about the resulting structure in different chromosomal regions (e.g., G- and R-bands).

CHROMOSOME BANDS AND MODELS OF THE CHROMOSOME SCAFFOLD

If transcription complexes play a very direct role in chromosome condensation at metaphase and in the final form and shape of the metaphase chromosome and attachments to its scaffold[79] then, given what we already know about the distribution of genes along the chromosome (chapter 3), we would predict that within such a model mitotic scaffold attachments would be by far the most frequent in T- and R-bands and the least frequent in G- and C-bands. Hence loops in the T-band regions of the genome would be several fold shorter than those found in G-band regions. Loops in regions of heterochromatin might be expected to be enormous within this scenario and the loop size of gene-rich chromosomes such as HSA19 would be very different from those on the gene poor chromosomes, e.g., HSA18 (Fig. 4.2b). Given that metaphase chromosomes are superficially cylinders of fairly uniform diameter, except at primary and secondary constrictions, there are no morphological features in the human karyotype that indicate that this model could be correct. In chapter 5 we discuss whether transcription complexes or their remnants are present on metaphase chromosomes is discussed.

If anything, this view of the chromosome is the opposite scenario to that suggested by observations of the path of SARs through chromosomes in which G-bands are predicted to have loop sizes that are shorter or approximately the same size as R-bands (Fig. 4.2a).[60] It is not clear whether we know enough about the sequence characteristics of SARs to be able to spot them within long stretches of primary DNA sequence. If this is possible then the higher order structure of the chromosome may be able to be decoded from the emerging sequence information of the genome.

Drawing on the strong visual characteristics of the mitotic scaffold, fluorescence in situ hybridization (FISH) has been used to see (literally) what sort of DNA sequences associate with the axial core of metaphase chromosomes.[81] Loops of DNA that encompass ~300 to 500 kb of sequence can be detected and specific and consistent sites along the DNA are found at the bases of the loops on both sister chromatids in association with the axial core of the chromosome. It is reasonable to assume that such sites correspond to the approximate positions of attachments to the mitotic scaffold. These attachments were observed to correlate well with the position of chromosomal origins of replication, but did not seem to relate to patterns of transcription (Fig. 4.2c,d). For example, constitutive heterochromatin (at sites other than centromeres), where no transcription has been documented, appeared to be as frequently associated with the chromosome core as did euchromatin. Conversely, within the very highly transcribed rDNA on the human acrocentric chromosomes the associations with the mitotic scaffold appeared to map within the nontranscribed spacer (NTS) sequences rather than within the transcription unit (Fig. 4.3). The region of the NTS most closely in opposition to the chromosome core was that known to correspond with the chromosomal replication origin consistent with the model in Figure 4.2c and d.[81]

The idea that the sites on the chromosome that are used as replication origins are influenced by the topological constraints of the chromosome is supported by several other lines of evidence. In both the budding and fission yeasts origins of replication on the chromosome also act as sequences conferring autonomous replication on plasmids (ARSs). As in SARs, ARSs in yeast are isolatable.[82] In other eukaryotes the ARS assay is not able to identify origins of replication that are used in the context of the chromosome—indicative of the fact that DNA sequence alone is not sufficient for origin specification. The frequency at which DNA replication is initiated along the chromosome is generally characteristic for an organism and is directly related to the loop size for that organism.[83] In addition, where the frequency of origins changes within an organism in response to programmed development the loop size in the genome changes concomitantly. Perhaps most strikingly, as origins become less frequent and loops become bigger during *Xenopus* embryogenesis so the metaphase chromosomes

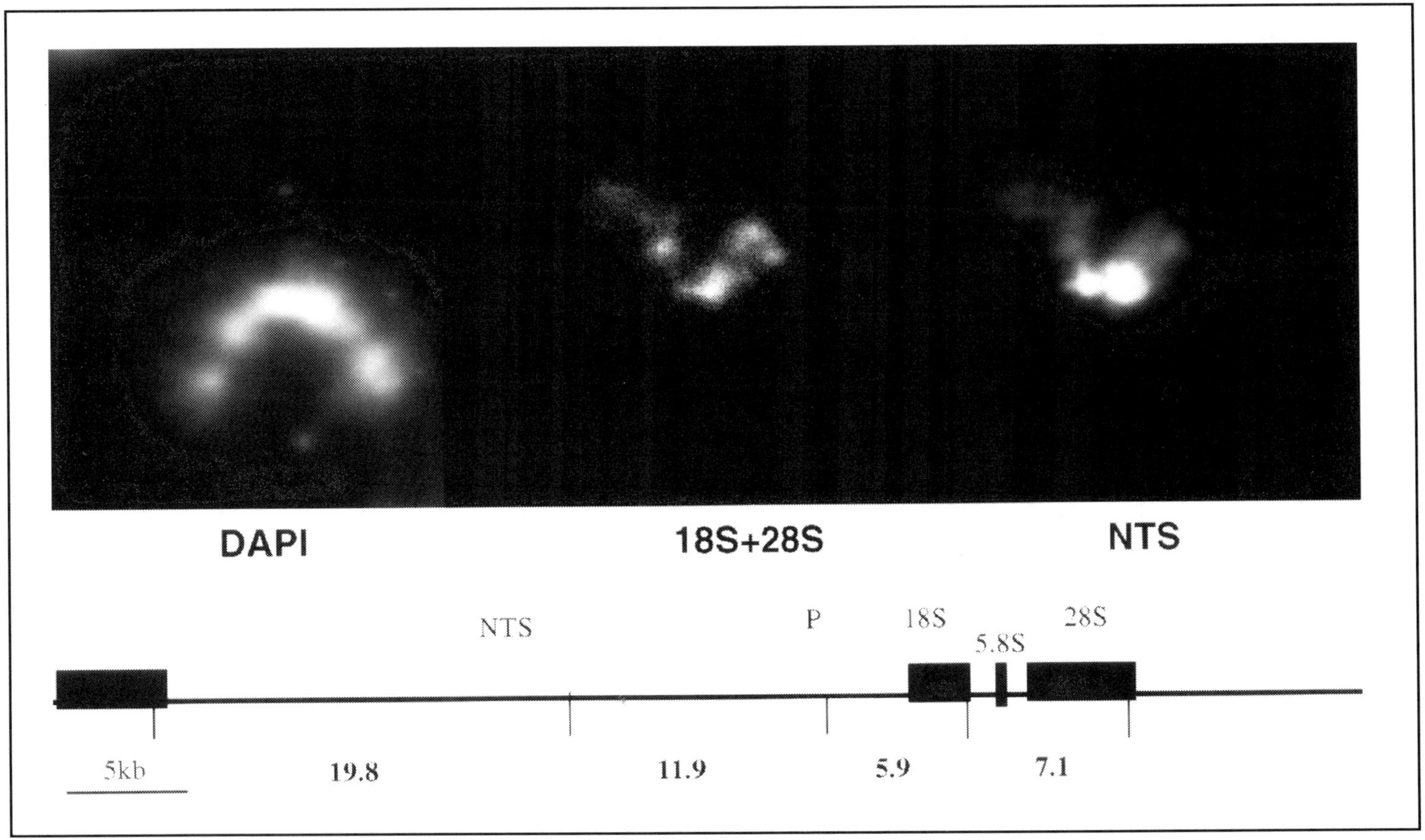

Fig. 4.3. The nontranscribed spacer (NTS) of the human rDNA repeat is associated with the chromosome core. A human metaphase acrocentric chromosome extracted with high salt and hybridized by FISH with probes for the 18 and 28S rRNA genes and for the NTS. Signal from the NTS, but not from the rRNA genes concentrates at the chromosome core.[81] The map of the human rDNA repeat unit is shown below the FISH analysis. Replication initiation begins in the 11.9 kb Eco RI fragment 5′ of the 18S gene.[94]

become shorter and fatter as would be predicted from the version of the radial loop/scaffold model depicted in Figure 4.1c and d.[84] In mammalian cells, known chromosomal origins have been shown to be associated with the chromosome scaffold or matrix throughout the cell cycle[78] and the in vivo sites of scaffold-associated topoisomerase II cleavage at interphase and metaphase comap with origins.[85] It is presumed that replicon size, and hence the spacing of replication origins, is the same in euchromatin and heterochromatin and between G-and R-bands. Hence we might expect loop sizes to be fairly uniform along the length of the metaphase chromosome and the resulting chromosomes to be cylinders of fairly uniform diameter in this model (Fig. 4.1c,d).

SPECIALIZED REGIONS OF THE METAPHASE CHROMOSOME SCAFFOLD

Indeed, chromosomes do generally appear to be of uniform width along their length, but there are a few exceptions. The most distinctive morphological landmark of mammalian metaphase chromosomes is the primary constriction—the site of the centromere. This constriction suggests that centromeric chromatin is in a higher order conformation that is different from that in the chromosome arms. In *S. cerevisiae,* sequences involved in conferring centromere activity are, like ARSs, scaffold-associated.[82] Sequences conferring centromere activity in mammals have yet to be functionally defined but the alphoid satellite is found at all human centromeres and visually, it is intimately associated with the mitotic scaffold with no large loops of DNA visible.[81] It is tempting to imagine that this arrangement of the DNA sequences at centromeres contributes not only to the physical appearance of the centromeric regions of metaphase chromosomes but also to their function.

This type of higher-order organization for centromeres was first suggested by the morphology of the centromeres of newt meiotic prophase lampbrush chromosomes which are marked by smooth round chromomeres with no associated lateral loops.[86] Similar to the tight association of human alphoid DNA with the core of mitotic chromosomes, there is a concentration of centromeric minor satellite DNA (that like human alphoid DNA contains a CenP-B binding motif) over the synaptonemal complex of rodent pachytene chromosomes rather than in the loops.[87] There is therefore evidence for specialized higher-order folding of centromeric DNA in chromosomes both at mitosis and meiosis.

We have already seen that there is an abundance of the major scaffold proteins topoisomerase II and ScII at centromeres. Minor protein components of the mitotic scaffold include specialized proteins, e.g., the INCENPs (chapter 1) and CenP-B, C and E that are part of the kinetochore.[88-90] The retention of these proteins at the scaffold of extracted mitotic chromosomes may contribute both to the morphology of the residual kinetochore and to the observed distribution of centromeric DNA within the core of the chromosome. The unusual chromatin structure of the centromere is further discussed in chapter 6.

THE CHROMOSOME SCAFFOLD VS. THE NUCLEAR MATRIX

The metaphase scaffold is a structure for which seeing is believing. However, extraction of interphase nuclei with some of the same agents used to reveal the mitotic scaffold leaves behind a residual proteinaceous structure that has also been called the nuclear scaffold/matrix or skeleton, dependent on the extraction method used. Sequences left behind in nuclei after extraction under both high salt and physiological ionic conditions have been related to both the processes of replication and transcription. SARs left behind after extraction with the detergent-like molecule lithium diiodosalicylate (LIS) do not appear to have any relationship with transcriptional activity. There is a lengthy and often contradictory literature on the existence, nature and functions of a scaffold- or matrix-like structure within nuclei. Here we will only consider what relationship such an interphase structure might have with the metaphase scaffold—intuition tells us that there must be at least some things in common if chromosomes are to refold in essentially the same way at each cell division and if they are to carry heritable epigenetic messages in their structure.

The residual protein content of 2M salt-treated chromosomes and nuclei have been compared.[91] Interphase nuclei have a much more complex pattern of remaining proteins, including the presence of lamins, than do metaphase chromosomes. However, some

DNA sequences defined as SARs in interphase nuclei can also be isolated as SARs at metaphase.[70] The colocalization of topoisomerase II with the chromosome and nuclear scaffold has also been used as a way of identifying the DNA sequences located at the bases of loops. Treating cells with compounds that trap the cleavable complex of topoisomerase II leave double strand breaks in the DNA at the sites of in vivo topoisomerase II action. These studies have identified the same cut sites both in interphase nuclei and in metaphase cells but not all cutting sites corresponded with the position of SARs.[85] Other studies have also concluded that the average loop size is the same in metaphase and interphase cells and figures anywhere from 15 to 650 kb have been quoted depending on the method used.[92,93] This similarity might not exclude the possibility that at least some attachments to the scaffold are dynamic through the cell cycle.

An appealing idea is that attachments to the mitotic scaffold set up the basic groundstate of chromatin structure that is propagated from one cell to its daughters and confers some important information to daughter cells about how to handle the genomes that they inherit. During interphase other temporary sites may be recruited to a scaffold or matrix through a variety of biochemical activities including transcription and replication. How can a function of the scaffold as a central architectural element of chromosomes directing the partitioning of the chromatin fiber into loop domains be married with the idea of a place where active and dynamic processes in the nucleus occur? Recent ideas that tackle this conundrum include those that suggest that at interphase a matrix is formed that serves as a network of channels on the surface of chromosomes that connects with the cytoplasm through nuclear pores and through which compounds necessary for the transcription and replication of the genome are guided in, and through which the products of the genome (mRNA) are exported.[93] It is not inconceivable that, in addition to a centrally located chromosome scaffold that is important for both chromosome shape and some functional activities of the chromosome at metaphase and interphase, there is an additional nuclear matrix that interacts with chromosomes (probably at their surface) only during interphase.

Table 4.1. General proteins concentrated in particular band types of mammalian metaphase chromosomes

T-Bands	R-Bands	G-bands	C-bands	Ref.
Tri-and tetra-acetylated H4	Tri-and tetra-acetylated H4			11
			Unacetylated H4	18
		H1		29
		HMGI/Y		42
		Topoisomerase II	Topoisomerase II	59,60
			Topoisomerase I	64
			ScII	49

Only the chromosome bands which show the highest concentration of protein by immunofluorescence at metaphase or by immunoprecipitation are shown. This does not preclude the presence of these proteins at other chromosomal sites at lower concentrations.

In this chapter we have discussed the fundamental differences in chromatin structure between different flavors of euchromatin (Table 4.1) and the implications that this has for the chromosome biology of G- and R-bands. In chapter 6 we discuss further the extreme case of heterochromatin and the special classes of proteins found associated with this form of chromosome band. In the next chapter we consider how the organization of dynamic aspects of chromosome function such as replication, transcription and recombination relate to chromosome banding patterns and to the concentrations of particular chromosomal proteins to regions of different band flavor that we have just discussed.

REFERENCES

1. Sumner AT. Chromosome Banding. London: Unwin Hyman, 1990.
2. Lee DY, Hayes JJ, Pruss D and Wolffe AP. A positive role for histone acetylation in transcription factor access to nucleosomal DNA. Cell 1993; 72:73-84.
3. Vettesse-Dadey M, Walter P, Chen H et al. Role of the histone amino termini in facilitating binding of a transcription factor, GAL4-AH, to nucleosome cores. Mol. Cell Biol. 1994; 14:970-981.
4. Wolffe A. Chromatin: Structure and Function. Academic Press, 1995.
5. Krude T. Nucleosome assembly during DNA replication. Curr. Biol. 1995; 5:1232-1234.

6. Ling X, Harkness TA, Schultz MC et al. Yeast histone H3 and H4 aminotermini are important for nucleosome assembly in vivo and in vitro: redundant and position-independent functions in assembly but not in gene regulation. Genes Develop. 1996; 10:686-699.
7. Sobel RE, Cook RG, Perry CA et al. Conservation of deposition-related acetylation sites in newly synthesized histones H3 and H4. Proc. Natl. Acad. Sci. USA 1995; 93:1237-1241.
8. Turner BM and O'Neill LP. Histone acetylation in chromatin and chromosomes. Seminars Cell Biol. 1995; 6:229-236.
9. Brownell JE, Zhou J, Ranalli T et al. Tetrahymena histone acetyltransferase A: a homolog to yeast Gcn5p linking histone acetylation to gene activation. Cell 1996; 84:843-851.
10. Taunton J, Hassig CA and Schreiber SL. A mammalian histone deacetylase related to the yeast transcriptional regulator Rpd3p. Science 1996; 272:408-411.
11. Jeppesen P and Turner BM. The inactive X chromosome in female mammals is distinguished by a lack of histone H4 acetylation, a cytogenetic marker for gene expression. Cell 1993; 74:281-289.
12. Houben A, Belyaev ND, Turner BM and Schubert I. Differential immunostaining of plant chromosomes by antibodies recognizing acetylated histone H4 variants. Chrom. Res. 1996; 4:191-194.
13. D'Anna JA, Gurley LR and Tobey RA. Extent of histone modifications and H1 content during cell cycle progression in the presence of butyrate. Exp. Cell Res. 1983; 147:407-417.
14. Turner BM. Acetylation and deacetylation of histone H4 continue through metaphase with depletion of more-acetylated isoforms and altered site usage. Exp. Cell Res. 1989; 182:206-214.
15. Tazi J and Bird A. Alternative chromatin structure at CpG islands. Cell 1990; 60:909-920.
16. Hebbes TR, Clayton AL, Thorne AW and Crane-Robinson C. Core histone hyperacetylation co-maps with generalized DDase I sensitivity in the chicken globin chromosomal domain. EMBO J. 1994; 13:1823-1830.
17. Turner BM, Franchi L and Wallace H. Islands of acetylated histone H4 in polytene chromosomes and their relationship to chromatin packaging and transcriptional activity. J. Cell Sci. 1990; 96:335-346.
18. O'Neill LP and Turner BM. Histone H4 acetylation distinguishes coding regions of the human genome from heterochromatin in a differentiation-dependent but transcription-independent manner. EMBO J. 1995; 14:3946-3957.
19. Mann RK and Grunstein M. Histone H3 N-terminal mutations allow hyperactivation of the yeast GAL 1 gene in vivo. EMBO J. 1992; 11:3297-3306.

20. Covault J and Chalkey R. Identification of distinct populations of acetylated histones. J. Biol. Chem. 1980; 255:9110-9116.
21. Nickel BE, Alli DC and Davie JR. Ubiquitinated histone H2B is preferentially located in transcriptionally active chromatin. Biochem. 1989; 28:958-963.
22. Matsui SI, Seon BK and Sandberg AA. Disappearance of a structural chromatin protein A24 in mitosis: Implications for molecular basis of chromatin condensation. Proc. Natl. Acad. Sci USA 1979; 76:6386-6390.
23. Golderer G and Grobner P. ADP-ribosylation of core histones and their acetylated subspecies. Biochem. J. 1991; 277:607-610.
24. Mahadevan LC, Willis AC and Barratt MJ. Rapid histone H3 phosphorylation in response to growth factors, phorbol esters, okadaic acid and protein synthesis inhibitors. Cell 1991; 65:775-783.
25. Sullivan KF, Hechenberger M and Masri K. Human CenP-A contains a histone H3 related histone fold domain that is required for targetting to the centromere. J. Cell Biol. 1994; 127:581-592.
26. Postnikov YV, Shick VV, Belyavsky AV et al. Distribution of high mobility group proteins 1/2, E and 14/17 and linker histones H1 and H5 on transcribed and non-transcribed regions of chicken erythrocyte chromatin. Nucl. Acids Res. 1991; 19:717-725.
27. Ball DJ, Gross DS and Garrard WT. 5-Methylcytosine is localized in nucleosomes that contain histone H1. Proc. Natl. Acad. Sci. USA 1983; 80:5490-5494.
28. Ridsdale JA, Hendzel MF, Decluve GP and Davie JR. Histone acetylation alters the capacity of the H1 histones to condense transcriptionally active competent chromatin. J. Biol. Chem. 1990; 265:5150-5156.
29. Breneman JW, Yau P, Teplitz RL and Bradbury EM. A light microscope study of linker histone distribution in rat metaphase chromosomes and interphase nuclei. Exp. Cell Res. 1993; 206:16-26.
30. Bustin M, Yamasaki H, Goldblatt D et al. Histone distribution in chromosomes revealed by antihistone sera. Exp. Cell Res. 1976; 97:440-444.
31. Th'ng JPM, Wright PS, Hamaguchi J et al. The FT210 cell line is a mouse G2 phase mutant with a temperature-sensitive cdc2 gene product. Cell 1990; 63:313-324.
32. Paulson JR, Patzlaff JS and Vallis AJ. Evidence that the endogeneous histone H1 phosphatase in HeLa mitotic chromosomes is protein phosphatase 1, not protein phosphatase 2. J. Cell Sci. 1996; 109:1437-1447.
33. Paulson JR. Sulphydryl reagents prevent dephosphorylation and proteolysis of histones in isolated HeLa metaphase chromosomes. Eur. J. Biochem. 1980; 111:189-197.

34. Lin R, Cook RG and Allis CD. Proteolytic removal of core histone amino termini and dephosphorylation of histone H1 correlate with the formation of condensed chromatin and transcriptional silencing during *Tetrahymena* macronuclear development. Genes Develop. 1991; 5:1601-1610.
35. Roth SY and Allis CD. Chromatin condensation: does histone H1 dephosphorylation play a role? Trends Biochem. Sci. 1992; 17:93-98.
36. Ohsumi K, Katagiri C and Kishimoto T. Chromosome condensation in Xenopus mitotic extracts without histone H1. Science 1993; 262:2033-2035.
37. Shen X, Yu L, Weir JW and Gorovsky MA. Linker histones are not essential and affect chromatin condensation in vivo. Cell 1995; 82:47-56.
38. Grosschedl R, Giese K and Pagel J. HMG domain proteins: architecural elements in the assembly of nucleoprotein structures. Trends Genet. 1994; 10:94-100.
39. Smith BJ, Robertson D, Birbeck MSC et al. Immunochemical studies of high mobility group non-histone chromatin proteins HMG1 and HMG 2. Exp. Cell Res. 1978; 115:420-423.
40. Ner SS and Travers AA. HMG-D, the Drosophila melanogaster homologue of HMG-1 protein, is associated with early embryonic chromatin in the absence of histone H1. EMBO J. 1994; 13:1817-1822.
41. Friedmann M, Holth LT, Zoghbi HY and Reeves R. Organization, inducible-expression and chromosome localization of the human HMGI(Y) nonhistone protein gene. Nucl. Acids Res. 1993; 21:4259-4267.
42. Disney JE, Johnson KR, Magnuson NS et al. High-Mobility Group protein HMG-I localizes to G/Q- and C-bands of human and mouse chromosomes. J. Cell Biol. 1989; 109:1975-1982.
43. Thanos D, Du W and Maniatis T. The high mobility group protein HMGI (Y) is an essential structural component of a virus-inducible enhancer complex. Cold Spring Harbor Quant. Biol. 1993; LVIII:73-81.
44. Paulson JR and Laemmli UK. The structure of histone-depleted metaphase chromosomes. Cell 1977; 12:817-828.
45. Marsden MPF and Laemmli UK. Metaphase chromosome structure: evidence for a radial loop model. Cell 1979; 17:849-858.
46. Howell WM and Hsu TC. Chromosome core structure revealed by silver-staining. Chromosoma 1979; 73:61-66.
47. Lewis C and Laemmli U. Higher order metaphase chromosome structure: evidence for metalloprotein interactions. Cell 1982; 29:171-181.

48. Earnshaw WC and Heck MMS. Localization of topoisomerase II in mitotic chromosomes. J. Cell Biol. 1985; 100:1716-1725.
49. Saitoh N, Goldberg IG, Wood ER and Earnshaw WC. ScII: An abundant chromosome scaffold protein is a member of a family of putative ATPases with an unusual predicted tertiary structure. J. Cell Biol. 1994; 127:303-318.
50. Hirano T and Mitchison TJ. Topoisomerase II does not play a scaffolding role in the organization of metaphase chromosomes assembled in Xenopus egg extracts. J. Cell Biol. 1993; 120:601-612.
51. Hirano T and Mitchison TJ. A heterodimeric coiled-coil protein required for mitotic chromosome condensation in vitro. Cell 1994; 79:449-458.
52. Yang L, Wold MS, Li JJ et al. Roles of DNA topoisomerases in simian virus 40 DNA replication in vitro. Proc. Natl. Acad. Sci. USA 1987; 84:950-954.
53. Uemura T, Ohkura H, Adachi Y et al. DNA topoisomerase II is required for condensation and separation of mitotic chromosomes in *S. pombe.* Cell 1987; 50:917-925.
54. Shamu CE and Murray AW. Sister chromatid separation in frog extracts requires DNA topoisomerase II activity during anaphase. J. Cell Biol. 1992; 117:921-934.
55. Adachi Y, Luke M and Laemmli UK. Chromosome assembly in vitro: topisomerase II is required for condensation. Cell 1991; 64:137-148.
56. Swedlow JR, Sedat JW and Agard DA. Multiple chromosomal populations of topoisomerase II detected in vivo by time-lapse, three-dimensional wide-field microscopy. Cell 1993; 73:97-108.
57. Earnshaw WC, Halligan B, Cooke CA et al. Topoisomerase II is a structural component of mitotic chromosome scaffolds. J. Cell Biol. 1985; 100:1706-1715.
58. Gasser SM, Laroche J, Falquet J et al. Metaphase chromosome structure. Involvement of topoisomerase II. J. Mol. Biol. 1986; 118:613-629.
59. Sumner AT. The distribution of topoisomerase II on mammalian chromosomes. Chromosome Res. 1996; 4:5-14.
60. Saitoh Y and Laemmli UK. Metaphase chromosome structure: bands arise from a differential folding path of the highly AT-rich scaffold. Cell 1994; 76:609-622.
61. Watt PM and Hickson ID. Structure and function of type II DNA topoisomerases Biochem. J. 1994; 303:681-695.
62. Vassetzky YS, Dang Q, Benedetti P and Gasser SM. Topoisomerase II forms multimers in vitro: effects of metals, β-glycerophosphate and phosphorylation of its C-terminal domain. Mol. Cell. Biol. 1994; 14:6962-6974.

63. Dang Q, Alghisi GC and Gasser SM. Phosphorylation of the C-terminal domain of topoisomerase II by casein kinase II affects DNA-protein interaction. J. Mol. Biol. 1994; 243:10-24.
64. Maul GG, French BT, van Venrooiji WJ and Jimenez SA. Topoisomerase I identified by Scleroderma 70 antisera: enrichment of topoisomerase I at the centromere in mouse mitotic cells before anaphase. Proc. Natl. Acad. Sci. USA 1986; 83:5145-5149.
65. Strunnikov AV, Larionov VL and Koshland D. SMC1: an essential yeast gene encoding a putative head-rod-tail protein is required for nuclear division and defines a new ubiquitous protein family. J. Cell Biol. 1993; 123:1635-1648.
66. Strunnikov AV, Hogan E and Koshland D. SMC2, a Saccharomyces cerevisiae gene essential for chromosome segregation and condensation defines a subgroup within the SMC family. Genes Dev. 1995; in press.
67. Saka Y, Sutani T, Yamashita Y et al. Fission yeast cut3 and cut14, members of a ubiquitous protein family, are required for chromosome condensation and segregation in mitosis. EMBO J. 1994; 13:4938-4952.
68. Gasser SM. Coiling up chromosomes Curr. Biol. 1995; 5:357-360.
69. Ma X, Saitoh N and Curtis PJ. Purification and characterization of a nuclear DNA-binding factor complex containing topoisomerase II and chromosome scaffold protein 2. J. Biol. Chem. 1993; 268:6182-6188.
70. Mirkovitch J, Gasser SM and Laemmli UK. Scaffold attachment of DNA loops in metaphase chromosomes. J. Mol. Biol. 1988; 200:101-110.
71. Gasser SM, Amati BB, Cardonas ME and Hofmann J F-X. Studies on scaffold attachment sites and their relation to genome function. Intl. Rev. Cytol. 1989; 119:57-96.
72. Adachi Y, Kas E and Laemmli UK. Preferential, cooperative binding of DNA topoisomerase II to scaffold-associated regions. EMBO J. 1989; 8:3997-4006.
73. Gimenez-Abian JF, Clarke DJ, Mullinger AM et al. A postprophase topoisomerase II-dependent chromatid core separation step in the formation of metaphase chromosomes. J. Cell Biol. 1995; 131:7-17.
74. Boy de la Tour E and Laemmli UK. The metaphase scaffold is helically folded: sister chromatids have predominantly opposite helical handedness. Cell 1988; 55:937-944.
75. Izaurralde E, Käs E and Laemmli UK. Highly preferential nucleation of histone H1 assembly on scaffold-associated regions. J. Mol. Biol. 1989; 210:573-585.
76. Zhao K, Kas E, Gonzalez E and Laemmli UK. SAR-dependent mobilization of histone H1 by HMGI/Y in vitro: HMGI/Y is en-

riched in H1-depleted chromatin. EMBO J. 1993; 12:3237-3247.
77. Strick R and Laemmli UK. SARs are cis DNA elements of chromosome dynamics: synthesis of a SAR repressor protein. Cell 1995; 83:1137-1148.
78. Dijkwel PA, Wenink PW and Poddighe J. Permanent attachment of replication origins to the nuclear matrix in BHK cells. Nucl. Acids Res. 1986; 14:3241-3249.
79. Cook PR. RNA polymerase: structural determinant of the chromatin loop and the chromosome. BioEssays 1994; 16:425-430.
80. Hendzel MJ, Sun J-M, Chen HY et al. Histone acetyltransferase is associated with the nuclear matrix. J. Biol. Chem. 199; 269: 22894-22901.
81. Bickmore WA and Oghene K. Visualizing the spatial relationship between defined DNA sequences and the axial core of extracted chromosomes. Cell 1996; 84:95-104.
82. Amati BB and Gasser SM. Chromosomal ARS and CEN elements bind specifically to the yeast nuclear scaffold. Cell 1988; 54:967-978.
83. Buongiorno-Nardelli M, Micheli G, Carri MT and Marilley M. A relationship between replicon size and supercoiled loop domains in the eukaryotic genome. Nature 1982; 298:100-102.
84. Micheli G, Luzzatto AR, Carri MT et al. Chromosome length and DNA loop size during early embryonic development of *Xenopus laevis*. Chromosoma 1993; 102:478-483.
85. Razin SV, Hancock R, Iarovaia O et al. Structural-functional organization of chromosomal domains. Cold Spring Harbor Symp. Quant. Biol. 1993; 58:25-35.
86. Callan HG. Lampbrush chromosomes. Proc. R. Soc. London B. 1982; 214:417-448.
87. Moens PB and Pearlman RE. Telomere and centromere DNA are associated with the cores of meiotic prophase chromosomes. Chromosoma 1990; 100:8-14.
88. Cooke CA, Bernat RL and Earnshaw WC. CenpB: a major human centromere protein located beneath the kinetochore. J. Cell Biol. 1990; 110:1475-1488.
89. Tomkiel JE, Cooke CA, Saitoh H et al. CenP-C is required for maintaining proper kinetochore size and for a timely transition to anaphase. J. Cell Biol. 1994; 125:531-545.
90. Yen TJ, Compton DA, Earnshaw WC and Cleveland DW. CenP-E, a human centromere associated protein released from chromosomes at the onset of anaphase. EMBO J. 1991; 10:1245-1991.
91. Lewis C, Lebkowski JS, Daly AK and Laemmli UK. Interphase nuclear matrix and metaphase scaffolding structures. J. Cell Sci. 1984; Suppl. 1:103-122.

92. Jackson DA, Dickinson P and Cook PR. The size of chromatin loops in HeLa cells. EMBO J. 1990; 9: 567-571.
93. Razin SV and Gromova II. The channels model of nuclear matrix structure. BioEssays 1995; 17:443-450.
94. Yoon Y, Sanchez JA, Brun C and Huberman JA. Mapping of replication initiation sites in human ribosomal DNA by nascent-strand abundance analysis. Mol. Cell. Biol. 1995; 15:2482-2489.

CHAPTER 5

The Activity of Chromosome Bands at Interphase

In this chapter we will consider how information about the major events that take place in and on chromosomes during interphase of the cell cycle can be related to the banding patterns evident on the same chromosomes at metaphase when these same activities have ceased. We will begin with the two major nucleic acid polymerization events that occur on chromosomes: the duplication of the DNA during S phase and the transcription of genes into RNA. We will also address the exchange of genetic material between chromatids or chromosomes and the repair of DNA damage. Lastly, we will consider how chromosomes leave interphase and become progressively condensed through prophase to metaphase. In some cells of some organisms, for example the larval salivary glands of *Drosophila*, individual interphase chromosomes can be easily seen since the homologous chromosomes are paired and endoreduplicated (polytenized). These interphase polytene chromosomes can be stained with dyes to reveal a series of bands along their length (Fig. 5.1) that superficially resemble the chromosome bands of mammalian metaphase chromosomes. However, the relationship between polytene bands and those of metaphase chromosomes has not been extensively studied and will not be substantially considered further here. For our purposes we can assume that individual interphase chromosomes are not readily discernible, but we will consider how metaphase chromosome bands can be reporters of interphase activities of the chromosome.

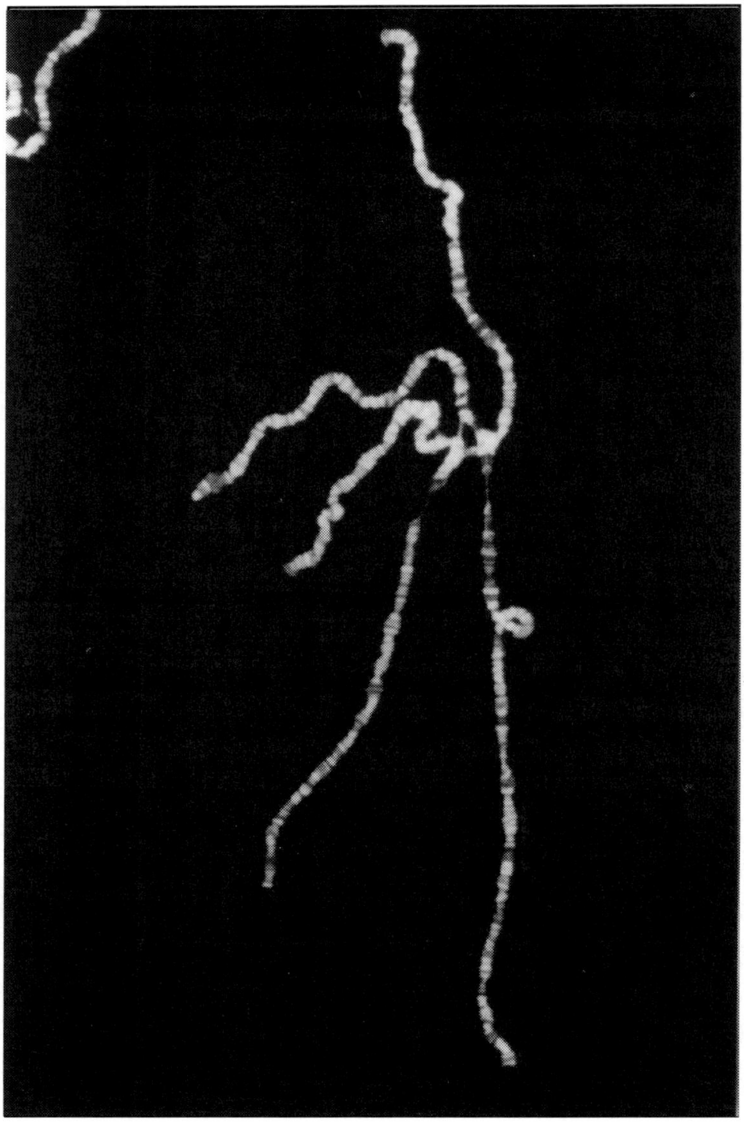

Fig. 5.1. Interphase bands on polytene chromosomes. Polytene chromosomes from larval salivary glands of the Canton S strain of Drosophila melanogaster *stained with DAPI. Note that the four chromosomes are joined at the chromocenter. Each chromosome is in fact an homologous pair that has been replicated many times in the absence of cell division so that many DNA molecules are aligned together. Courtesy of Margarette Heck, Edinburgh.*

DNA REPLICATION—WHERE TO BEGIN

Chromosomal replication is initiated bidirectionally from sites on chromosomes termed replication origins. The rate of progression of the replication fork outward from these origins is approximately 2-3 kb of DNA/minute in eukaryotes. Hence, if the replication of an average human chromosome were initiated from a single centrally located origin it would take 17 days and nights to be completed—clearly not a practical proposition. Fortunately, the replication of eukaryotic chromosomes is initiated from multiple origins spaced along chromosomes, each replication unit being termed a replicon. Our archetypal human chromosome probably uses on the order of 1500 origins in order that replication is completed within the eight hours or so of a typical S phase.

In man replicons are thought to be 50 kb to 600 kb in size[1] and it has been suggested that clusters of replicons may represent individual high resolution chromosome bands.[2] A high resolution chromosome band (~5 Mb) might be replicated from around 50 origins whose use is coordinately controlled. The clustering of replication origins is evident in interphase nuclei where in vivo labeling has shown that the sites of ongoing DNA replication occur in visible 'factories' that also contain many of the proteins (e.g., DNA polymerase α, PCNA, RP-A, DNA methyltransferase, cyclin A and CDK2) needed for the replication of chromatin.[3-6] It is tempting to imagine that all the origins in one high resolution chromosome band are located in the same replication factory at interphase and this is a hypothesis that should be testable in the not too distant future. However, since there are thought to be only a few hundred factories, but several thousand high-resolution chromosome bands[7] there must be more than one chromosome band per factory and individual factories might even contain origins originating from more than one chromosome. For example, do the origins of replication for the rDNA loci on the five human acrocentric chromosomes all colocalize in the same factory?

What is the initiating event for the formation of replication factories and what marks out a replication origin? Both could be established anew during G1, or there could be some aspect of the forerunners of factories and origins brought into the nucleus on the metaphase chromosomes.

In most eukaryotes replication origins do not seem to be specified solely by particular sequence motifs but rather by aspects of chromatin structure and chromosome context. Indeed, the sites where replication initiates can change during development in response to remodeling of chromosomes.[8] Since replication origins appear to be heritable from one cell to another some facet of them must be encoded within the structure of metaphase chromosomes. Tantalizingly, replication factories are associated with the nuclear scaffold,[5] at least one component of prereplicative complexes is present on mitotic chromosomes[3] and replication origins themselves are associated with the nuclear scaffold throughout the cell cycle, including at metaphase.[9] Indeed, replication origins may adopt a distinctive position within the interior of metaphase chromosomes close to the chromosome core.[10] Hence the organization of DNA into loops constrained upon the chromosome scaffold may play a central role in determining replicon frequency in eukaryotes. This means that replicon frequency and loop size are two fundamental periodicities in the genome that go hand in hand in an organism and that may vary together during development.[11,12] Conversely, the remnants of replication factories may play a role in directing the formation of metaphase chromosomes (Fig. 4.2).

In *S. cerevisiae* where origins (ARSs—for autonomously replicating sequences) are specified by short DNA sequences, the proteins of the ORC (origin recognition complex) and ABF-1 (ARS binding factor), required for initiation of replication, are stably bound at origins throughout the cell cycle, including mitosis. Late in mitosis this interaction is modified to allow the origin to be used.[13] In mammals, the correct replication origins may not be available for use until later in G1.[14] Although ARSs have been shown to act as SARs in *S. cerevisiae*,[15] no interaction between ORC and the chromosome scaffold has yet been identified.

DNA REPLICATION—WHEN TO BEGIN

Chromosome banding techniques that rely upon the incorporation of BrdU into early or late replicating DNA provide the most graphic illustration that the time of replication of different regions of eukaryotic chromosomes is strictly controlled and intimately related to chromosome organization (chapter 2). Gene-rich T- and

R-band compartments of the genome are replicated early in S phase while the gene-poor euchromatin of G-bands is replicated in the latter half of S, a feature that is exploited in replication banding (Fig. 2.3). A more careful analysis of replication banding patterns shows an even tighter association between chromosome band flavor and replication time. The relative replication times of euchromatic bands have been characterized cytogenetically as occupying one of eighteen time spans, (with I the earliest and XVIII the latest) using protocols of pulse chase with BrdU.[16] The time at which a chromosome band replicates is related to its response to G-banding, to the frequency of Alu repeats that it contains and to its gene density and base composition (Table 5.1). Hence T-bands replicate on average earlier (stage 4) than ordinary R-bands (stage 6), with G-bands following on behind (stage 12 on average). There are no autosomal R-bands that have failed to complete their replication by stage 10, whereas no G-bands start replicating before this time. Hence, the two broad categories of euchromatic chromosome band (G- and R-) are treated quite distinctly by the

Table 5.1. A gradation of properties with chromosome band flavor

	G_{Cond}	G_{Dark}	G_{All}	R_{All}	T	T_{Alu}
Proportion of euchromatin (%)	8.3	22.6	45.5	54.4	20.2	13.4
Proportion of genes (%)	2.4	9.1	20	80	45.6	35.5
Observed/expected gene density	0.29	0.4	0.44	1.47	2.26	2.63
Mean replication time (I-18)	11.8	13.1	12.0	5.4	3.6	3.2
Fusion index (from 1250 to 300 bands)	0	1.02	1.51	1.79	1.84	2.08

Band types are of increasing R-band flavor from left to right and this is indicated by the shading. G_{Cond} are G-bands that are already fully condensed by prophase. G_{Dark} are the darkest staining G bands. G_{All} and R_{All} are all G- or R-bands, respectively. T_{Alu} are the most Alu-rich T-bands. These categorizations are based upon the analyses in refs. 45 and 89. Average replication time was as determined cytogenetically[16] and the fusion index was calculated from the analysis in ref. 69.

nuclear replication machinery. The relative replication timing determined cytologically by BrdU incorporation for human chromosome 11, in relation to other chromosome banding procedures is illustrated in Figure 2.5. This temporal pattern of replication for human chromosome 11 determined by gross cytological analysis is confirmed by a more detailed cytological FISH-based analysis of the relative replication time of individual chromosome 11 loci from different bands (Fig. 5.2).[17]

What aspects of the differences in primary DNA sequence or chromatin structure between chromosome bands of different flavors could be responsible for the link with the temporal control of DNA replication? In addition, or alternatively, could differences in replication time be determining some of the other characteristics of chromosome bands? One of the most influential features that differs between chromosome band types may be gene density. It has been suggested that sites of transcription at the G1/S boundary seed the assembly of the first replication factories in early S phase. Hence, the most transcriptionally active regions of the genome, that would also correspond statistically to the regions with the highest concentrations of genes, would tend to be the ones to be replicated first.[18]

Relationships between transcription and replication timing have been studied at a molecular level. In mammals most housekeeping genes appear to replicate relatively early in S phase. Tissue-specific genes fall into two main categories: those that always replicate at the same stage of S phase (whether early or late), regardless of their state of expression, and those that replicate earlier when transcribed than when inactive.[19,20] Hence, the time of replication of a DNA sequence is not absolutely fixed—it can be influenced by other nuclear activities and by chromosome context. In addition to individual genes, whole chromosome bands,[21] or even whole chromosomes[22] can switch their replication time under certain circumstances. The mainly transcriptionally inactive X chromosome, although late replicating, still replicates its R-bands before its G-bands, and before some autosomal G-bands.[23] This observation suggests that relative replication timing is, at least to some extent, a fundamental process related to the chromosome banding pattern

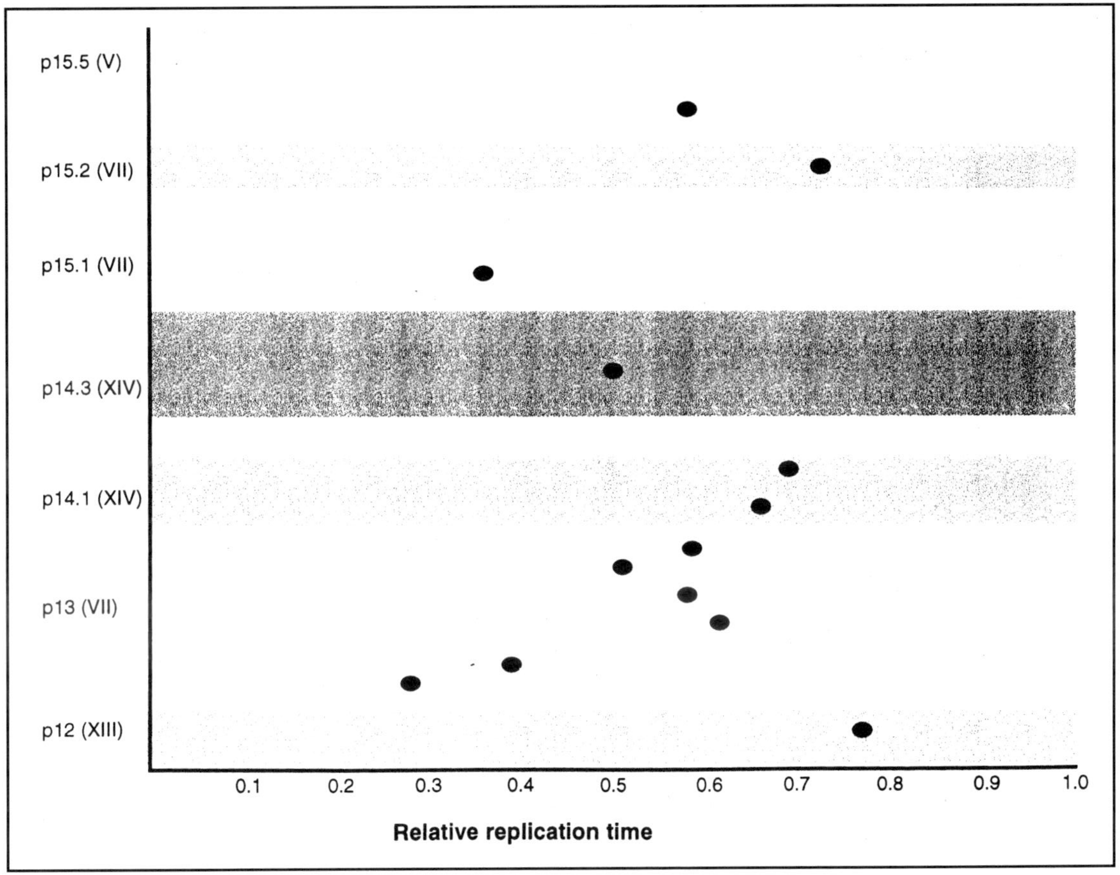

Fig. 5.2. Timing of replication of loci on 11p assayed by cytogenetic location and by FISH analysis. The cytogenetic replication time (Roman numerals) of each chromosome band was taken from ref. 16. Data on FISH analysis were taken from ref. 17 and from unpublished data. The time of replication as a proportion of S phase was determined by FISH for individual loci as described in ref. 17. 0 is replication at the start of S phase and 1 is the end of S phase. All loci in G-bands (shaded grey) were determined to replicate in the second half of S phase, no R-band locus replicated later than 60% of the way through S. However, some individual G-band loci replicated earlier than individual R-band loci.

that is not completely dependent on the transcriptional status of genes on the chromosome. Epigenetic modifications of chromatin that appear to modulate replication timing in mammals include DNA methylation and histone acetylation.[17]

DNA that is replicated late in S phase (e.g., G-band DNA) is not passively replicated by the passage of replication forks that have been initiated earlier. In *S. cerevisiae* it has been shown that there are barriers on the chromosome to prevent replication forks from early regions passing into late replicating DNA.[24] Instead, late replicated DNA is copied from origins that do not fire until late in S phase and replication factories that are formed in late S phase are distinct from early replication factories.[18,25] What prevents the late factories from forming, and a late origin of replication from firing, at the beginning of S? Is it inherent to some property of the DNA sequence itself or is it dependent on some aspect of the chromosome context or the time of replication in the previous cell cycle? The emerging answer from studies with the well characterized origins of *S. cerevisiae* seems to be a bit of both. Early ARSs can be converted to late firing ones by moving them to some other chromosomal sites such as close to telomeric sequences. Conversely, a late replicating origin can fire early when removed from its normal chromosome context and placed on an extrachromosomal plasmid.[26] However, some ARSs seem to be inherently late replicating no matter where you put them on the chromosome[27] and sequence determinants for late replication are located outside of the ARS itself.[28]

There are many more potential origins on the chromosome than are ever used, and while this might seem a good insurance policy to ensure that the entire genome gets replicated, it makes the business of policing replication to ensure that no sequence is copied more than one per cell cycle appear more difficult. Within yeast rDNA arrays, although there is a potential origin in each repeat unit, only the origin from every third to tenth unit appears to get used in any one particular cycle[29] and use of one origin seems to preclude or interfere with the use of other potential origins located close by.[27] The mechanism of this interference is not understood but it may operate through the topological constraints imposed by the association of origins with a chromosome or nuclear scaffold. Sequestering of origins into the chromosome core[10] once they have been used could be imagined as a way to help prevent their reuse within the same cell cycle.

THE CONSEQUENCES OF REPLICATION TIMING

Whereas Q-, G- and R-banding patterns have only been observed in some eukaryotes, replication banding is almost universal among living organisms possessing chromosomes large enough to visualize by microscopy, suggesting that it is a fundamental consequence of, or requirement for, the compartmentalization of complex genomes.

There are models of chromatin structure that invoke replication time as a means of regulating gene expression[22] through the window of opportunity to remodel chromatin that is afforded by the process of chromatid duplication. An attractive idea is that replication during the early stages of S phase exposes DNA to chromosomal proteins that tend to result in open, transcriptionally competent chromatin that in turn replicates early in the next cell cycle. Conversely, replication during late S phase might take place in an environment, or even a particular nuclear compartment, dominated more by chromosomal proteins that tend to lock chromatin into an inactive state. However, direct evidence to support this idea is lacking. In fact, it has been shown recently in *S. cerevisiae* that newly-replicated copies of transcriptionally-active ribosomal RNA gene are regularly packed into nucleosomes and that regeneration of the active chromatin structure occurs only some time after replication by a mechanism that involves disruption of nucleosomes.[30]

In chapter 3 we discussed how differences in replication timing might have accounted for the accumulation of recently retroposed SINEs into R-bands and LINEs into G-bands. Gene density and base composition may also be related mechanistically, through synchronous replication of chromosome domains in factories at which the composition of free dNTPs available for ongoing synthesis varies during S phase.[31] This might occur either through spatial or temporal changes in dNTP pools. If levels of free dGTP or dCTP are at their highest at factories formed during early S phase, then over time the earliest replicated DNA in the germ line would evolve to have an elevated G + C base composition as compared with the DNA replicated late in S phase. This idea explains the existence of isochores without having to invoke an argument of selection to explain how they got there. If this is the case then would expect to find isochores lacking in those organisms who best maintain a homogeneous dNTP pool throughout the nucleus and throughout S phase. Is this a reasonable proposition? In the nucleus there are thought to be only enough free dNTPs, especially dGTP, for a few minutes (i.e., <10 kb) of replication.[32] The activity and substrate specificity of ribonucleotide reductase involved in the supply of dNTPs is subject to feedback control from the concentration of the different nucleotides excepting dCTP. Instead, dCTP acts allosterically on dCMP deaminase, an enzyme whose activity controls the relative amounts of dCTP and dTTP. It is known that experimentally-induced biases in the dNTP pools increases mutation rates. This supports the idea that the concentration of free nucleotides can influence the bases inadvertently incorporated into the genome in error.

It is not clear whether variations in base composition along yeast chromosomes are completely analogous to isochores (the variations in base composition in the yeast genome do not extend to intergenic DNA), but peaks of high GC do correspond with regions of early replication in at least some cases.[33] Even in bacteria there appears to be a link between replication timing and base composition.[34]

TRANSCRIPTION

Transcription and replication can be considered to be very similar enzyme driven processes that act upon DNA. Both involve recognizing a place to start on the chromosome, unwinding of the DNA helix and polymerization of a complementary nucleic acid strand from the DNA template. Like replication, transcription is thought to occur in nuclear factories and RNA polymerases have been found on the nucleoskeleton at interphase.[18] It has even been suggested that the remnants of transcription factories aggregate at mitosis and play a role in the formation of the condensed mitotic chromosome (Fig. 4.2).[35] If this were so it would be hard to understand what the basis for the condensation of genetically inert constitutive heterochromatin would be since there are presumably no transcription factories associated with it. Other predictions that would be made by such a model are that a chromosome, such as human chromosome 19, that has a very high density of genes might be expected to have many more transcription factories associated with it than a chromosome such as HSA18 with approximately 5-fold fewer genes on it (Fig. 3.6) and consequently, that it would have significantly different dimensions at metaphase. The same lines of argument would predict that T-bands would have different physical parameters than would G-bands.

What is the fate of transcription, transcription factories and transcribed RNA at mitosis? There is general agreement that there is no active transcription by RNA polymerase I or II at mitosis[36] and that nascent RNAs are aborted as cells enter mitosis.[37] However, the polymerases themselves can still be found on the chromosomes at appropriate locations and there are some suggestions that transcription can be reactivated merely by decondensation of metaphase chromosomes.[38,39] While RNA polymerases I and II can be identified as components of the yeast nuclear scaffold, RNA polymerase II is not a component of the mammalian mitotic scaffold.[40,41]

The fate of transcription factors at mitosis is variable. For some polymerase II and III transcription factors their phosphorylation at mitosis inhibits their DNA binding activity and they are lost from mitotic chromosomes.[42] Other factors are displaced through unknown mechanisms while some are even retained on the chromosome. The latter class includes the RNA polymerase I transcription factor UBF.[43]

Although transcription may have a limited role in determining the structure of chromosomes at metaphase it is clear that metaphase chromosomes must contain messages that can influence patterns of gene expression in daughter cells. Consistent with this, DNase I hypersensitive sites in promoters are retained in mitotic chromatin as are the distinctive chromatin structures of active and inactive rRNA genes.[42,44]

RECOMBINATION AND REPAIR

Recombination occurs most often between sister chromatids in mitotically dividing cells and between homologous chromosomes at meiosis. Variations in the density of meiotic chiasmata along mammalian chromosomes are apparent when physical and genetic maps are compared. T-bands have the highest rates of exchange, followed by "mundane" R-bands, then G-bands. Heterochromatin shows the lowest rates of recombination.[45] Hence, the recombination frequency is highest in the most GC rich isochore H3 and is low in L1 and L2.[46] For example, on human chromosome 9, 52% of the genetic map is found in 20% of the euchromatin in the region of two R-bands, one of which is a T-band.[47] Similarly on chromosome 21, 33% of the genetic map is found in 9% of the euchromatin—also in the region of a T-band.[48] Irrespective of band flavor, there is a general trend for mouse and human chromosomes to show high recombination frequencies near telomeres and low frequencies across centromeres.[49,50]

Why should meiotic recombination be influenced by chromosome band flavor? G- and R-bands behave differently during meiotic pairing, synapses and crossing over.[51] R-bands are both the sites of synaptic initiation and the preferred sites of crossing over in mammals and other vertebrates.[52] This may reflect differences between bands in transcription, replication, base composition or chromatin structure in premeiotic nuclei. For example, the earlier a band replicates the longer its relative synaptonemal length at meiosis.[53,54] A recombination hot spot in the mouse MHC locus was found to colocalize with binding sites for two transcription factors and with a strong nuclease hypersensitive site.[55]

In *S. cerevisiae* there is also a correlation between patterns of gene density and genetic recombination along the entire length of chromosome III but not between recombination and base composition[56,33] and sites for meiosis-induced double-stranded breaks comap with DNase I hypersensitive sites over the 5' end of open reading frames.[57] Thus recombination may merely initiate where DNA is accessible to recombination enzymes, as in promoter regions[58] and be excluded from tightly packed, genetically silent, heterochromatin.[59] Mutants in *Caenorhabditis elegans (C. elegans)* that disrupt the normal patterns and biases in recombination rates along chromosomes suggest a more proactive correlation between chromosomal environment and recombination rather than just a passive one.[60] In this organism most meiotic exchanges occur in the terminal portions of the chromosomes, but these are gene-poor regions.[61]

How does the condensed structure of the chromosome at meiosis compare to its mitotic counterpart? Within the synaptonemal complex between homologous pairs of meiotic prophase chromosomes an axial core, to which loops of chromatin are anchored, can be seen. We do not know whether this axial core is analogous to either the interphase nuclear matrix or to the mitotic chromosome scaffold—it does not seem to contain SARs.[62] However, like the mitotic scaffold it can be used as a substrate on which to perform high resolution FISH and the association of particular parts of the genome with this axial core can also be shown.[63]

The rate of DNA repair is also thought to be dependent on transcriptional activity, time of replication and chromatin structure and hence likely to be highest in T-bands and conversely, DNA repair linked replication may influence base composition.[34,46,64]

Similarly, in mammalian cells forms of somatic mitotic recombination are enhanced by increases in transcription and by CpG demethylation.[65,66] Levels of repair are highest in actively transcribed genes and this makes sense since it would be more deleterious if mistakes made in these regions were perpetuated than would be the case for errors in gene-free regions. Biochemically the link between repair and gene density is maintained by the coupling of repair enzymes to the transcription machinery.[67]

CHROMOSOME CONDENSATION

Once the genome has been transcribed, replicated and errors repaired it is time to condense once again to allow for chromosome segregation. Eukaryotic chromosomes progressively condense to different degrees as they enter prophase and proceed through prometaphase to metaphase. As well as being necessary to protect the physical integrity of the chromosome during the dynamics of mitosis, condensation may play a direct role in allowing sister chromatids to be decatenated and segregated from one another.[68] Condensation is not uniform along chromosome length and, for certain types of chromosome band, can be delayed or inhibited completely. Early studies of the morphological steps of chromosomal condensation from early to late prophase found that late replicating regions (G- and C-bands) condense early in prophase and the early replicating regions (R- and T-bands) condense late in prophase. The most comprehensive study of chromosome condensation followed chromosomes as they condensed from prophase (1,250 band resolution) to late metaphase (300 bands).[69] Using both R-banding and G-banding techniques, to counteract any optical bias (the phenomenon of a band looking larger or smaller depending on the banding technique), the process of band "fusion" was observed and the fusion index of different bands was calculated. The fusion index is defined as the average number of fusions per sub-band required to go from a given band stage to the 300 band stage. It is calculated by counting all post-decimal digits in the ISCN nomenclature, for each 300-resolution-equivalent band, at the higher resolution and dividing by the number of higher resolution sub-bands. An average euchromatic sub-band undergoes a mean of 1.59 fusions while progressing from prophase to late metaphase, but about this mean R-sub-bands fuse more readily with each other than do G-bands. Indeed, by prophase many G-bands have already attained maximal contraction and fuse no further as mitosis progresses, whereas R-bands still have much contraction (fusion) to go. For example, in Figure 5.3 compare the behavior of the T-band 11p15 (the most terminal band on the short arm at 350 band resolution) with the adjacent11p14 which is a G-band. These two bands appear to be of the same size by both G- and R-bands methods at a 350 band resolution (Fig. 2.5). At 550 band resolution p14 is still not subdivided whereas p15 consists of five sub-bands. 11p14 is comprised of three prophase sub-bands (1250 band resolution) but at this early stage p15 can be differentiated into nine sub-bands. Table 5.1 shows that this applies to the rest of the genome as well—as Giemsa bands get lighter and more gene-rich they become earlier replicating and their fusion index increases.

A consequence of this differential condensation is that R-band chromatin appears to occupy a progressively smaller proportion of euchromatin as metaphase approaches. In Table 5.2 the figures

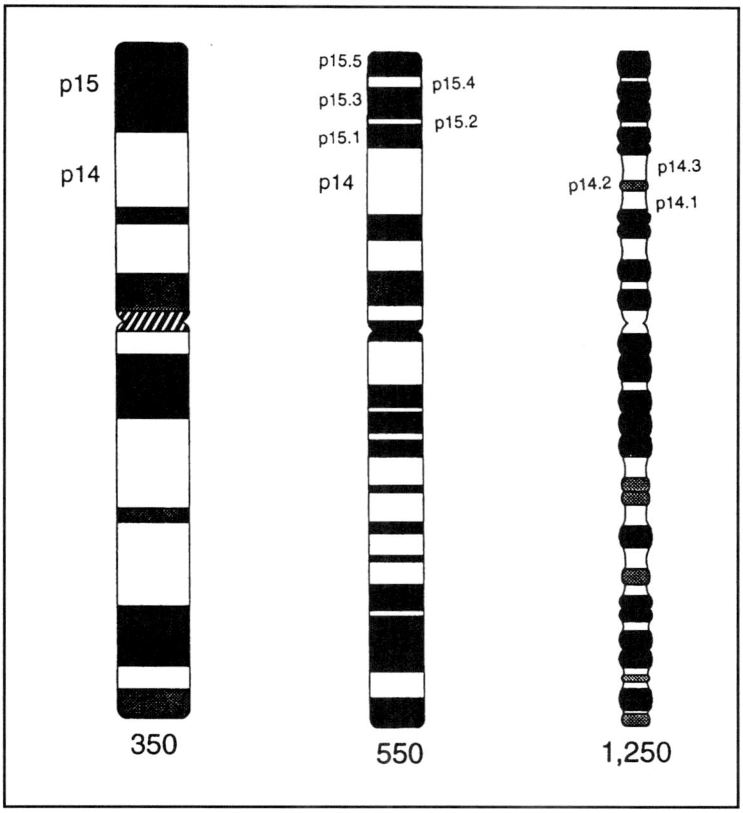

Fig. 5.3. Subdivisions of G- and R/T-bands. R-band ideograms are shown for chromosome 11s at 350 and 550 metaphase band resolution and 1250 prophase resolution. R-bands are shown in black. Note that the T-band 11p15 is comprised of more sub-bands than the G-band 11p14 and that the sub-bands of 11p15, as well as being more numerous, are also slower to fuse into one.

for the proportion of euchromatin occupied by the major band types are taken from 1250, 850, 400 and 300-band resolution ideograms.[70] Whereas R-band chromatin occupies around 70% of the karyotype at prophase (1250 bands) this decreases to 50% by late metaphase (300 bands).

How does fusion occur? Studies both by light microscopy and electron microscopy have concluded that sub-bands fuse by some chromatin reorganization or reshaping rather than by a simple longitudinal contraction.[71] This process is reversible mechanically since, rather dramatically, G-banded chromosomes can be stretched by up to five times their metaphase length to reveal sub-bands

Table 5.2. Comparison of euchromatic genome occupation for R- and G-bands from prophase (1250 bands) to metaphase (300 bands)

Resolution (band no)	Chromosome Occupancy of G-bands	Chromosome Occupancy of R-bands	Ref.
1250	29%	71%	70
850	46.7%	53.3%	89
400	45.5%	54.4%	90
300	51.4%	48.6%	90

G-bands account for an apparently greater proportion of the genome in late metaphase due to preferential condensation of R-band chromatin. Measurements for 1250 band resolution are given in ref. 70. All other measurements were taken directly from ideograms. It should be noted that the only ideograms based on actual measurements are the 850 and 1250-band ones, although the compilers of the ISCN ideograms did faithfully reproduce relative band lengths.

normally seen only at a 2,000 band resolution.[72] Differences in the condensation of chromosome bands are not artificially induced by the banding techniques themselves—they have been seen by studying the topology of fixed, unstained mitotic chromosomes by atomic force microscopy. Such chromosomes display variations in the thickness of the chromatid fiber corresponding to a G-banding pattern.[73]

Several lines of evidence suggest that we should look to chromosome coiling as a source of the differences in G- and R-band condensation. In mammalian chromosomes, the coils of the chromatid fiber have been observed to occasionally change their direction of gyre and intermediately-coiled chromosomes frequently showed local over-coiling in regions corresponding to G-bands.[74] There have been claims that during progressive trypsin treatment of chromosomes coils come to look like bands with coils closely packed in G-bands and more widely spaced in R-bands.[75]

The idea of differential chromosome coiling between bands is resurrected and expanded by recent observations and ideas that point to the mitotic chromosome scaffold as a central player in the process (chapter 4).[76] It is tempting to speculate that a scaffold protein such as ScII might direct the progressive coiling of the chromosome core as mitosis proceeds beginning first in G-bands. Are mitotic G- and R-bands then not separate entities but just morphologically levels of chromosome packaging captured at different stages along the same route? If so what is controlling the differential condensation of the chromosome in regions with different band flavor or is the sort of coiling within them fundamentally different (Fig. 4.1)?

Mitotic G-bands correspond to meiotic chromomeres. We do not know whether the differential packaging of pachytene chromosomes into chromomeres and inter-chromomeres reflects the same packaging differences that occur between mitotic G-bands and R-bands. We also have yet to determine whether there are differences between these band types during the various stages of interphase in somatic cells.

Mitotic condensation, especially that of heterochromatin, can be experimentally inhibited with a variety of agents including Hoechst, DAPI, distamycin A, 5-azacytidine and actinomycin D.[77] Dye-based inhibitors are effective during G2, just prior to metaphase, and there is mounting evidence that their target may be the mitotic scaffold. Depleting either of the two major scaffold proteins from a cell free extract interferes with normal chromosome condensation[78,79] as does adding in an artificial protein that should bind SARs strongly.[80]

Inhibition of condensation by the base analog 5-azacytidine requires the inhibitor to be present during S phase. It is an inhibitor of methyltransferases and covalently links the enzyme to the DNA. At low doses the sites of undercondensation correspond to heterochromatin, known to harbor the highest concentrations of 5-MeC (chapter 3). G-bands are inhibited from condensing next so that an R-band like pattern is seen on the resultant chromosomes. At extreme doses the chromosomes take on a pulverized appearance.[81]

There are a few unusual situations in which portions of mammalian chromosomes fail to condense correctly. The first of these are fragile sites—visible discontinuities or even breaks in the metaphase structure. Fragile sites have been mainly studied on human chromosomes and most of them are not revealed spontaneously but only after manipulation of culture conditions. These culture conditions give some clue as to the defect in chromosome condensation. Some fragile sites can be induced with distamycin

A or 5-azacytidine, two agents whose use in the inhibition of chromosome condensation through interference with the chromosome scaffold or through inhibition of methylation has already been discussed. Most other agents that induce fragile sites interfere with DNA replication, including aphidicolin—a specific inhibitor of DNA polymerase α, and Brdu (chapter 2).[82] The fragile sites that have attracted the most attention, and for which we can begin to understand their molecular basis, are the folate-sensitive sites that include FRAXA—the fragile site associated with fragile X mental retardation. The use of folate inhibitors or the omission of folic acid in culture results in depleted pools of dTTP or dCTP (chapter 2). Molecular analysis of FRAXA and a number of other folate-sensitive fragile sites revealed the presence of unstable $p(CGG)_n$ repeats in the underlying DNA sequence. The expansion of these repeats leads to their methylation, and it is this that correlates with the appearance cytologically of a fragile site.[83] Why should repeat expansion and methylation result in aberrant metaphase chromosome condensation? The normal allele at the FRAXA locus replicates late in S phase whereas its amplified and methylated counterpart is even further delayed in its replication.[84] We have already seen that normally late replicating DNA is in fact the first to be condensed as cells enter prophase (Table 5.1). It is possible that the expanded and methylated fragile sites are so delayed in their replication that they miss cell cycle signals in G2/M that control condensation altogether.

Another situation in which unusual chromosome condensation was seen was when long stretches of chromosomal DNA from *S. pombe* were integrated into mouse cells. The sites of integration were more constricted than the endogenous euchromatin but neither did they C-band, indicating that they were not analogous to sites of constitutive heterochromatin.[85] The yeast DNA replicated along with the G-bands of the mouse euchromatin and its methylation state did not correlate with its cytological appearance. The constriction was suggested to result from the frequent attachments of the yeast DNA to the chromosome scaffold, resulting in smaller loops of DNA at the site of integration than in the endogenous DNA, and this was translated into a thinner chromatid fiber through the type of mechanism shown in Figure 4.1c and d. A similar morphology has been observed in mammalian cells at the sites of integration of *S. cerevisiae* yeast artificial chromosomes (YACs) carrying segments of human DNA.[86,87] These observations suggest that there are inherently different periodicites in yeast and mammalian chromosomes and that these differences can be maintained and transmitted through cell divisions in a heterologous system indicative of a level of chromatin structure that can be faithfully replicated.

Interphase chromosomes can be made to condense into metaphase-like structures (prematurely condensed chromosomes—PCCs) by fusion of interphase and mitotic cells[88] or by treatment with phosphatase or phosphodiesterase inhibitors. Banding can also be induced on such chromosomes. If the interphase cells are in S phase at the time of condensation the regions that were actively engaged in replication appear as extended regions on the PCCs showing the effects that replication can impose on higher levels of chromatin structure.

References

1. Hand R. Eucaryotic DNA: Organization of the genome for replication. Cell 1978; 15:317-325.
2. Holmquist GP. Evolution of chromosome bands: Molecular ecology of noncoding DNA. J. Mol. Evol. 1989; 28:469-486.
3. Adachi Y and Laemmli UK. Identificationof nuclear pre-replication centres poised for DNA synthesis in Xenopus egg extracts; immunolocalization study of replication protein A. J. Cell Biol. 1992; 119:1-15.
4. Leonhardt H, Page AW, Weier H-U and Bestor TH. A DNA targetting sequence directs DNA methyltransferase to sites of DNA replication in mammalian nuclei. Cell 1992; 71:865-873.
5. Hozák P, Hassan AB, Jackson DA and Cook PR. Visualisation of replication factories attached to a nucleoskeleton. Cell 1993; 73:361-373.
6. Cardoso MC, Leonhardt H and Nadal-Ginard B. Reversal of terminal differentiation and control of DNA replication; cyclin A and cdk2 specifically localize at sub-nuclear sites of DNA replication. Cell 1993; 74:979-992.
7. Yunis JJ. Mid-prophase human chromosomes. The attainment of 2000 bands. Hum. Genet. 1981; 56:293-298.
8. Hyrien O, Maric C and Mechali M. Transition in specification of embryonic metazoan DNA replication origins. Science 1995; 270: 994-997.

9. Dijkwel PA, Wenink PW and Poddighe J. Permanent attachment of replication origins to the nuclear matrix in BHK cells. Nucl. Acids Res. 1986; 14:3241-3249.
10. Bickmore WA and Oghene K. Visualizing the spatial relationship between defined DNA sequences and the axial core of extracted chromosomes. Cell 1996; 84:95-104.
11. Buongiorno-Nardelli M, Micheli G, Carri MT and Marilley M. A relationship between replicon size and supercoiled loop domains in the eukaryotic genome. Nature 1982; 298:100-102.
12. Micheli G, Luzzatto AR, Carri MT et al. Chromosome length and DNA loop size during early embryonic development of *Xenopus laevis*. Chromosoma 1993; 102:478-483.
13. Diffley JFX, Cocker JH, Dowell SJ and Rowley A. Two steps in the assembly of complexes at yeast replication origins in vivo. Cell 1994; 78:303-316.
14. Wu J-R and Gilbert DM. A distinct G1 step required to specifiy the Chinese Hamster DHFR replication origin. Science 1996; 271:1270-1272.
15. Amati BB and Gasser SM. Chromosomal ARS and CEN elements bind specifically to the yeast nuclear scaffold. Cell 1988; 54:967-978.
16. Dutrillaux B, Couturier J, Richer C-L and Viegas-Péguinot E. Sequence of DNA replication in 277 R- and Q-bands of human chromosomes using a BrdU treatment. Chromosoma 1976; 58:51-61.
17. Bickmore WA and Carothers AD. Factors affecting the timing and imprinting of DNA replication on a mammalian human chromosome. J. Cell Sci. 1995; 108:2801-2809.
18. Hassan AB and Cook PR. Does transcription by RNA polymerase play a direct role in the initiation of replication? J. Cell Sci. 1994; 107:1381-1387.
19. Goldman MA, Holmquist GP, Gray MC et al. Replication timing of genes and middle repetitive sequences. Science 1984; 224: 686-692.
20. Hatton KS, Dhar V, Brown EH et al. Replication program of active and inactive multigene families in mammalian cells. Mol. Cell. Biol. 1988; 8:2149-2158.
21. Adolph S, Hameister H and Schildkraut CL. Molecular analysis of the aberrant replication banding pattern on chromosome 15 in murine T-cell lymphomas. Chromosoma 1992; 101:388-398.
22. Riggs AD and Pfeifer GP. X-chromosome inactivation and cell memory. Trends Genet. 1992; 8:169-174.
23. Drouin R, Lemieux N and Richer C-L. Analysis of DNA replication during S phase by means of dynamic chromosome banding at high resolution. Chromosoma 1990; 99:273-280.

24. Zhu J, Newlon CS and Huberman JA. Localization of a DNA replication origin and termination zone on chromosome III of Saccharomyces cerevisiae. Mol. Cell Biol. 1992; 12:4733-4741.
25. O'Keefe RT, Henderson SC and Spector DL. Dynamic organization of DNA replication in mammalian cell nuclei: spatially and temporally defined replication of chromosome-specific α-satellite sequences. J. Cell Biol. 1992; 116:1095-1110.
26. Ferguson BM and Fangman WL. A position effect on the time of replication origin activation in yeast. Cell 1992; 68:333-339.
27. Brewer BJ, Diller JD, Friedman KL et al. The topography of chromosome replication in yeast. Cold Spring Harbor Symp. Quant. Biol. 1993; 58:425-434.
28. Friedman KL, Raghuraman MK, Fangman WL and Brewer BJ. Analysis of the temporal program of replication initiation in yeast chromosomes. J. Cell Sci. 1995; 19 (S):51-58.
29. Linskens MHK and Huberman JA. Organization of replication of ribosomal DNA in Saccharomyces cerevisiae. Mol. Cell. Biol. 1988; 8:4927-4935.
30. Lucchini R and Sogo JM. Replication of transcriptionallly active chromatin. Nature 1995; 374:276-280.
31. Wolfe KH, Sharp PM and Li W-H. Mutation rates differ among regions of the mammalian genome. Nature 1989; 337:283-285.
32. Reichard P. Interactions between deoxyribonucleotide and DNA synthesis. Ann. Rev. Biochem. 1988; 57:349-374.
33. Sharp PM and Lloyd AT. Regional base composition variation along yeast chromosome III: evolution of primary chromosome structure. Nucl. Acids Res. 1993; 21:179-183.
34. Deschavanne P and Filipski J. Correlation of GC content with replication timing and repair mechanisms in weakly expressed *E. coli* genes. Nucl. Acids Res. 1995; 23: 1350-1353.
35. Cook PR. A chromomeric model for nuclear and chromosome structure. J. Cell Sci. 1995; 108:2927-2935.
36. Johnson TC and Holland JJ. Ribonucleic acid and protein synthesis in mitotic HeLa cells. J. Cell Biol. 1965; 27:565-574.
37. Shermoen AW and O'Farrell PH. Progression of the cell cycle through mitosis leads to abortion of nascent transcripts. Cell 1991; 67:303-310.
38. Matsui S-I, Weinfeld H and Sandberg AA. Quantitative conservation of chromatin-bound RNA polymerases I and II in mitosis. J. Cell Biol. 1979; 80:451-464.
39. Scheer U and Rose KM. Localization of RNA polymerase I in interphase cells and mitotic chromosomes by light and electron microscopic immunocytochemistry. Proc. Natl. Acad. Sci. USA 1984; 81:1431-1435.

40. Cardenas ME, Laroche T and Gasser SM. The composition and morphology of yeast nuclear scaffolds. J. Cell Sci. 1990; 96:439-450.
41. Lewis CD, Lebkowski JS, Daly AK and Laemmli UK. Interphase nuclear matrix and metaphase scaffolding structures. J. Cell Sci. 1984; 1 (S):103-122.
42. Martinez-Balbas MA, Dey A, Rabindran SK et al. Displacement of sequence-specific transcription factors from mitotic chromatin. Cell 1995; 83:29-38.
43. Roussel P, Andre C, Masson C. et al. Localization of the RNA polymerase I transcription factor hUBF during the cell cycle. J. Cell Sci. 1993; 104:327-337.
44. Conconi A, Widmer RM, Koller T and Sogo JM. Two different chromatin structures coexist in ribosomal RNA genes throughout the cell cycle. Cell 1989; 57:753-761.
45. Holmquist GP. Chromosome bands, their chromatin flavors, and their functional features. Am. J. Hum. Genet. 1992; 51:17-37.
46. Eyre-Walker A. Recombination and mammalian genome evolution. Proc. Roy. Soc. Lon. [Biol]. 1993; 252:237-243.
47. Povey S, Smith M, Haines J et al. Report on the first international workshop on chromosome 9. Ann. Hum. Genet. 1992; 56:167-221.
48. Chumakov I et al. Continuum of overlapping clones spanning the entire human chromosome 21q. Nature 1992; 359:380-387.
49. Laurie DA and Hulten MA. Further studies on chiasma distribution and interference in the human male. Ann. Hum. Genet. 1985; 49:203-214.
50. Nachman MW and Churchill GA. Heterogeneity in rates of recombination across the mouse genome. Genet. 1996; 142:537-548.
51. Ashley T. Prediction of mammalian synaptic and recombinational behavior of inversion heterozygotes based on mitotic breakpoint data and the possible evolutionary consequences. Genetica 1990; 83:1-7.
52. McQueen HA, Fantes J, Cross SH et al. CpG islands of chicken are concentrated on minichromosomes. Nature Genet. 1996; 12:321-324.
53. Chandley AC. A model for effective pairing and recombination at meiosis based on early replicating sites (R-bands) along chromosomes. Hum. Genet. 1986; 72:50-57.
54. Luciani M, Guichaoua MR, Cau P et al. Differential elongation of autosomal pachytene bivalents related to their DNA content in human spermatocytes. Chromosoma 1988; 97:19-25.
55. Shenkar R, Shen M and Arnheim N. DNase I hypersensitive sites and transcription factor-binding motifs within the mouse EB meiotic recombination hotspot. Mol. Cell. Biol. 1991; 11:1813-1819.
56. Oliver S et al. The complete DNA sequence of yeast chromosome III. Nature 1992; 357:38-46.

57. Wu T-C and Lichten M. Meiosis-induced double-strand break sites determined by yeast chromatin structure. Science 1994; 263: 515-518.
58. Goldway A, Sherman D, Zenvirth D. et al. A short chromosomal region with major roles in yeast chromosome *III* meiotic disjunction, recombination and double strand breaks. Genet. 1993; 133:159-169.
59. Egel R, Willer M and Nielson O. Unblocking of meiotic crossing-over between the silent mating-type cassettes of fission yeast, conditioned by the recessive mutant RIK1. Curr. Genet. 1989; 15:407-410.
60. Zetka M-C and Rose AM. Mutant *rec-1* eliminates the meiotic pattern of crossing over in *Caenorhabditis elegans.* Genet. 1995; 141:1339-1349.
61. Barnes TM, Kohara Y, Coulson A and Hekimi S. Meiotic recombination, noncoding DNA and genomic organization in *Caenorhabditis elegans.* Genetics 1995; 141:159-179.
62. Pearlman RE, Tsao N and Moens PB. Synaptonemal complexes from DNase-treated rat pachytene chromosomes contain (GT)n and LINE sequences, but no MARs/SARs. Genet. 1991; 130:865-872.
63. Moens PB and Pearlman RE. Telomere and centromere DNA are associated with the cores of meiotic prophase chromosomes. Chromosoma 1990; 100:8-14.
64. Filipski J. Why the rate of silent codon substitutions is variable within a vertebrate's genome. J. Theor. Biol. 1988; 134:159-164.
65. Hsieh C-L and Lieber MR. CpG methylated minichromosomes become inaccessible for V(D)J recombination after undergoing replication. EMBO J. 1992; 11:315-325.
66. Nickoloff JA. Transcription enhances intrachromosomal homologous recombination in mammalian cells. Mol. Cell. Biol. 1992; 12:5311-5318.
67. Downes CS, Ryan AJ and Johnson RT. Fine tuning of DNA repair in transcribed genes: mechanisms, prevalence and consequences. BioEssays 1993; 15:209-216.
68. Holm C. Coming undone: how to untangle a chromosome. Cell 1994; 77:955-957.
69. Drouin R, Lemieux N and Richer C-L. Chromosome condensation from prophase to late metaphase: Relationship to chromosome bands and their replication time. Cytogenet. Cell Genet. 1991; 57:91-99.
70. Drouin R, Holmquist GP and Richer C-L. High-resolution replication bands compared with morphologic G- and R-bands. In: Harris H and Hirschhorn K, eds. Advances in Human Genetics. Vol. 22. New York: Plenum Press, 1994:47-115.

71. Messier P-E, Drouin R and Richer C-L. Electron microscopy of gold-labeled human and equine chromosomes. J. Histochem. Cytochem. 1989; 37:1443-1447.
72. Claussen U, Mazur A and Rubtsov N. Chromosomes are highly elastic and can be stretched. Cytogenet. Cell Genet. 1994; 66:120-125.
73. Musio A, Mariani T, Frediani C et al. Longitudinal patterns similar to G-banding in untreated human chromosomes: evidence from atomic force microscopy. Chromosoma 1994; 103:225-229.
74. Goradia RY and Davis BK. Banding and spiralisation of human metaphase chromosomes. Hum. Genet. 1977; 36:155-160.
75. Takayama S. Configurational changes in chromatids from helical to banded structures. Chromosoma 1976; 56:47-54.
76. Saitoh Y and Laemmli UK. Metaphase chromosome structure: bands arise from a differential folding path of the highly AT-rich scaffold. Cell 1994; 76:609-622.
77. Rocchi A, di Castro M and Prantera G. Effects of DAPI on human leukocytes in vitro. Cytogenet. Cell Genet. 1979; 23:250-254.
78. Adachi Y, Luke M and Laemmli UK. Chromosome assembly in vitro: topisomerase II is required for condensation. Cell 1991; 64:137-148.
79. Hirano T and Mitchison TJ. A heterodimeric coiled-coil protein required for mitotic chromosome condensation in vitro. Cell 1994; 79:449-458.
80. Strick R and Laemmli UK. SARs are cis DNA elements of chromosome dynamics; synthesis of a SAR repressor protein. Cell 1995; 83: 1137-1148.
81. Schmid M, Haaf T and Grunert K. 5-azacytidine—induced undercondensations in human chromosomes. Hum. Genet. 1984 67:257-263.
82. Sutherland GR and Richards RI. The molecular basis of fragile sites in human chromosomes. Curr. Opinions Genet. Develop. 1995; 5:323-327.
83. Nancarrow JK, Kremer E, Holman K et al. Implications of FRA16A structure for the mechanism of chromosomal fragile site genesis. Science 1994; 264:1938-1941.
84. Hansen RS, Canfield TK, Lamb MM et al. Association of fragile X syndrome with delayed replication of the FMR1 gene. Cell 1993; 73:1403-1409.
85. McManus J, Perry P, Sumner AT et al. Unusual chromosome structure of fission yeast DNA in mouse cells. J. Cell Sci. 1994; 107:469-486.
86. Featherstone T and Huxley C. Extrachromosomal maintenance and amplification of yeast artificial chromosome DNA in mouse cells.

Genomics 1993; 17:267-278.
87. Larin Z, Fricker MD and Tyler-Smith C. De novo formation of several features of a centromere following introduction of a Y alphoid YAC into mammalian cells. Hum. Mol. Genet. 1994; 3:689-695.
88. Gollin SM, Wray W, Hanks SK et al. The ultrastructural organization of prematurely condensed chromosomes. J. Cell Sci. 1984; 1(S):203-221.
89. Francke U. Digitised and differentially shaded human chromosome ideograms for genomic applications. Cytogenet. Cell Genet. 1994; 65:206-219.
90. Harnden DG and Klinger HP, eds. ISCN An International System for Human Cytogenetic Nomenclature. Published in collaboration with Cytogenet. Cell Genet. Karger, Basel 1985:48-57.

CHAPTER 6

Heterochromatin and Other Extremes of Chromatin Flavor

Thus far we have concentrated on the molecular differences between different flavors of euchromatin, for example, between R- and G-bands. We have largely ignored forms of chromosome banding (such as C-banding) that highlight heterochromatic portions of the genome. However, although cytologically G-bands are classed as part of euchromatin, many of the molecular properties of G-bands are akin to those of heterochromatin. It seems unlikely to us that biochemically there is a completely strict dividing line between heterochromatin and euchromatin—we might consider the very gene-rich T-bands and gene-free constitutive heterochromatin as the opposite extremes of a continuum of chromatin flavors. We are probably more advanced in our understanding of the detailed structure of heterochromatin than for any other part of the genome, due to the availability of genetic screens for components of heterochromatin. In this chapter we consider the composition, structure and behavior of the different types of heterochromatin in anticipation that it will have much to tell us about euchromatin in the future. At the other end of the spectrum, we also look at other extremes of chromatin structure, such as the hyperactive X chromosome of *Drosophila* males. There are surely lessons about the majority to be learned by studying the extreme few.

Heitz[1] was the first to note that some segments of eukaryotic chromosomes could remain visible (i.e., did not decondense) after metaphase and coined the term heterochromatin to describe their different behavior. Since this initial definition was cytological, some have argued that only organisms with visible chromosomes can truly be said to have heterochromatin. However, it is generally accepted that we now encompass a more molecular definition of heterochromatin. This incorporates regions of chromosomes from organisms where the chromosomes are too small to be seen by light microscopy but which have other accepted properties of heterochromatin, i.e., they appear to be condensed, genetically inert and late replicating. It is these very organisms that have been the subject of the elegant genetics that have now led us to a very exciting and new level of understanding of chromatin structure, particularly with respect to the proteins involved in the formation and maintenance of heterochromatin.

Heterochromatin is traditionally classified into two types. The first of these—constitutive heterochromatin—is thought to be present throughout the life of an organism, whereas facultative heterochromatin is defined as being under developmental control. Our increased understanding of developmental biology and the molecules involved in the formation of these types of heterochromatin may demand a rethink of these simplistic definitions.

CONSTITUTIVE HETEROCHROMATIN

In man, constitutive heterochromatin is found in regions of the genome that are C-band positive (chapter 2) and that appear to be inaccessible to nucleases.[2,3] These sites include those in and around the centromere of each chromosome and on the long arm of the Y chromosome. The amount of heterochromatin at these sites can be very variable between individuals (highlighting its genetic inertness). In *Drosophila*—an organism that combines the assets of good chromosome cytology and genetics—the main sites of constitutive heterochromatin are found at the chromocenter, telomeres and the fourth/Y chromosome. At the molecular level these sites are characterized by the presence of middle and highly repetitive sequences, often in tandem arrays of satellites (chapter 3) and the very repetitive nature of the DNA at these sites may contribute to their chromatin structure per se. There appears to be an innate ability of some tandem repeats integrated into the genome to form heterochromatin de novo or to propagate nearby pre-existing heterochromatin.[4] In some cases this can manifest as new C-bands.[5] In the mouse it has been shown that the same stretch of DNA can appear cytologically as euchromatic C-band negative material or as positive C-bands (and hence heterochromatin?) dependent on copy number. The threshold for the detection of C-bands was approximately 100 copies of a 100 kb repeat, i.e., a total of 10 Mb of DNA.[6] It is not clear if this represents a transition in the chromatin structure into a C-band/heterochromatic state or whether it merely reflects the cytogenetic resolution limit for the detection of C-bands or heterochromatin. If the latter, there may well be other elements spread about the genome that can confer the properties of heterochromatin but that are cytogenetically invisible. However, it is also clear that not every high copy number stretch of repetitive DNA is capable of forming C-banded heterochromatin since most large and visible homogeneous staining regions (HSRs) of amplified DNA are C-band negative[7] and express genes contained within them.

It is possible that the folding and pairing of some tandemly repeated sequences can provide a DNA template that facilitates the building of a heterochromatic chromatin structure. In turn, constitutive heterochromatin can allow large arrays of repetitive DNA to accumulate because, as we shall discuss later, recombination is usually suppressed within it. A complex folding and pairing of repetitive DNA sequences within constitutive heterochromatin may explain why antibodies directed against triplex DNA can highlight C-bands.[8]

In mammals, the main heterochromatic sites contain members of the classical satellite DNA families (chapter 3) and there is little evidence that these satellites are interspersed with unique or lower copy number sequences. In *Drosophila,* the α-heterochromatin that is recognizable cytologically on mitotic chromosomes around the centromere was originally considered to consist only of satellite DNAs and be completely gene-free. The β-heterochromatin was thought to connect the α-heterochromatin to euchromatin and to be only cytologically distinctive on interphase polytene chromosomes (where unlike α-heterochromatin it is not under-represented). Hence, it had been considered that it should be classed as facultative heterochromatin by some. β-heterochromatin was also thought to consist of middle repetitive DNA sequences and to have a reasonable gene density.[9] However, it is now clear that no such clear distinction exists at the molecular level—with simple sequence satellite DNA being interspersed with stretches of more complex DNAs (both unique and moderately repetitive DNA, including transposable elements) throughout heterochromatin.[10,11]

No endogenous genes have been mapped to regions of C-banded constitutive heterochromatin in the human genome (the hatched areas of Fig. 3.5). In addition, restriction analysis of mammalian genomic DNA indicates an absence of CpG islands in these same regions.[12] Systematic analysis of CpG island density on human chromosomes by FISH suggests that there are stretches of many megabases of DNA that are barren of CpG islands and that contain highly repetitive DNA elements, at most centromeres, on the long arm of the Y chromosome, on the short arms of the five acrocentric chromosomes and in the pericentric C-band material on chromosomes 1, 9 and 16 (Fig. 6.1). However, some eukaryotic genes do actually seem to prefer a heterochromatic location and cease to function when moved to euchromatic locations.[11,13-15] This suggests that these genes utilize the proteins of heterochromatic complexes to aid their transcriptional regulation. We might consider that these are extreme versions of genes that have adapted to life in G-bands and that, similarly, G-band genes may fail to function correctly when moved to other locations on chromosomes such as to R-band euchromatin.

FACULTATIVE HETEROCHROMATIN

Facultative heterochromatin is chromatin that is under developmental control—it starts out life as normal run of the mill euchromatin; it contains normal coding DNA and unlike constitutive heterochromatin it is not full of tandem repeats. It is subsequently inactivated and condensed in a developmentally regulated manner. In the most spectacular situations, entire chromosomes can become heterochromatinized and subsequently eliminated from cells during embryogenesis in some organisms.[16-18]

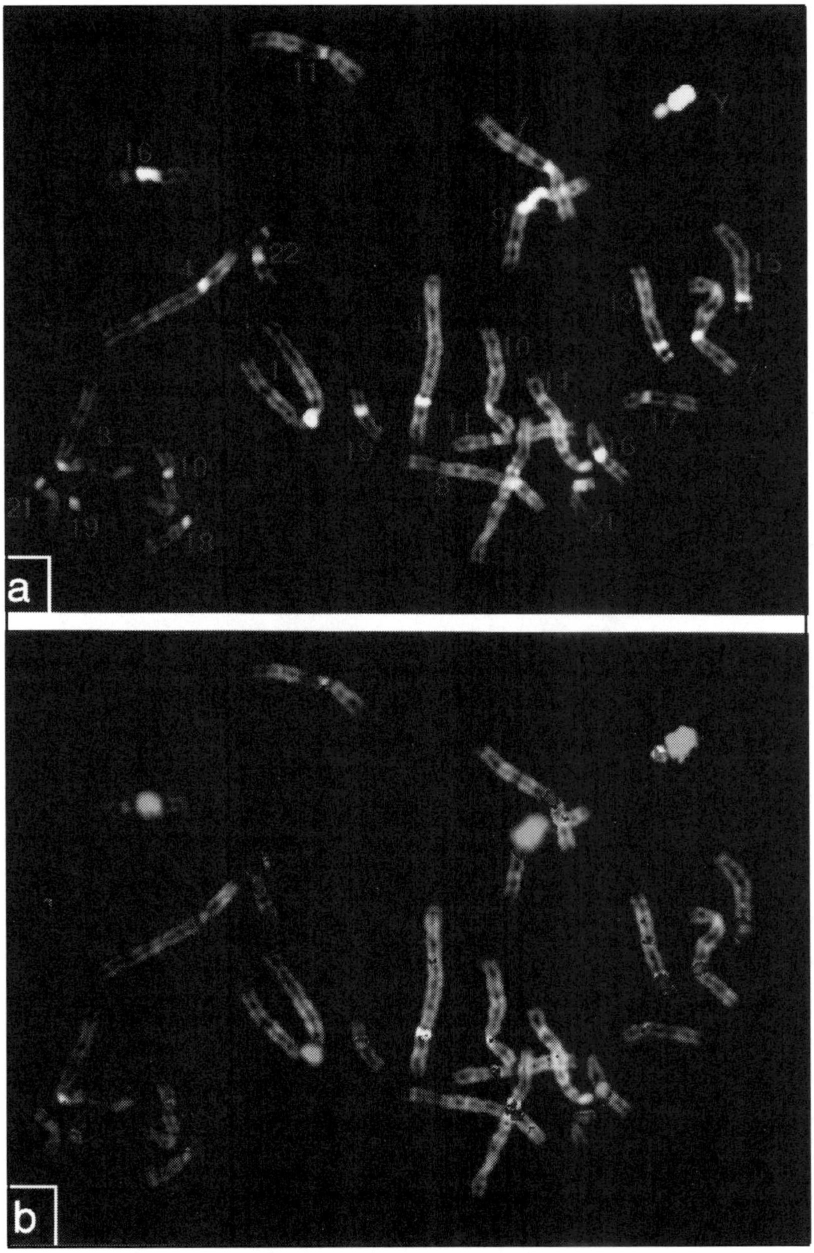

Fig. 6.1. Regions of the human genome barren of CpG islands are concentrated in regions of constitutive heterochromatin. (a) Human chromosomes G-banded with DAPI and actinomycin D. The C-band regions of the genome are stained the brightest. (b) The same chromosomes hybridized with a fraction of the human genome in which CpG islands are separated by greater than three megabases of DNA (green). Suppression with CotI DNA was reduced in this experiment so that signal from the repetitive sequences in C-bands could be seen.

In mammals the paradigm for facultative heterochromatin is the inactive X chromosome in female somatic cells that ensures dosage compensation of X-linked genes between the sexes. Despite having identical primary DNA sequences, the inactive and active X chromosomes can be distinguished by several banding techniques including modified G-banding and Q-banding (an illustration that fluorochrome banding is not solely a manifestation of underlying base composition), but the inactive X does not C-band.[19,20]

Additionally, an almost[21] universal feature of the inactive X in all mammals is that it is late replicating compared to its active counterpart and hence is distinctive in its replication banding. However, the replication of the inactive X is not just uniformly late. It has some early replicating bands (e.g., p11.2 and q13) and it replicates its R-bands before its G-bands.[22,23] In interphase human cells the inactive X chromosome is also usually considered to be identifiable as the condensed Barr body located at the periphery of the nucleus. Some evidence suggests that the X chromosome of the Barr body is bent around so that the telomeres are in close proximity to each other and that this configuration may be maintained to some extent in metaphase.[24] The chromosome is prematurely condensed at prophase and prometaphase compared to the rest of the karyotype and it is likely that this condensed chromatin state is responsible for the differences in behavior of the inactive and active Xs in response to some banding techniques.

The genetics and cytology of X chromosome inactivation is enabling us to disentangle the events that take place in the transition from euchromatin to facultative heterochromatin. X chromosome inactivation is controlled in cis from a genetically defined locus—the X-inactivation center (XIC, or *Xic* in the mouse).[25] The XIC is thought to control the counting mechanism that determines how many X chromosomes must be inactivated and the choice of X chromosome for activity as well as disseminating the inactivation signal along the chromosome. The XIC is now known to be a gene—XIST/*Xist*—that, uniquely, is transcribed before inactivation occurs only from the chromosome that is to become inactivated.[26] What is fascinating about this gene is that it does not encode a protein and that its long (17 kb) RNA product associates along the length of the inactive X chromosome (Barr body) itself in interphase.[27] How the transcript stays in association with the chromosome from which it was transcribed and how it spreads along the chromosome is not known, nor do we yet understand how XIST transcription begins the series of epigenetic changes that result in the repression of most, but not all, genes on that chromosome.

Although it has been clearly demonstrated that a functional XIST gene is required to initiate X-chromosome inactivation[26] its continued presence is not necessary to maintain the inactive state.[28] The inactivated state is stable through cell division, whereas XIST RNA does not stay associated with the inactive X-chromosome at anaphase.[27] Nevertheless, the inactive X is morphologically distinctive at prophase/metaphase[19,20] and, as we shall see in subsequent sections, at least some of the changes that follow on from XIST expression are changes to chromatin structure that are heritable through mitosis.

Another strategy that is used to achieve dosage compensation of X-linked genes is not to designate one X of a pair as almost completely inactive and heterochromatic, but to reduce transcription from both Xs in XX animals. This is the route chosen in *C. elegans.* One of the proteins, DPY-27, known to be involved in this process is discussed in the following section. We know little about the changes in the cytological properties and chromatin structure that accompany this dosage compensation process. Also, it seems unlikely that the repressed X chromosomes are heterochromatic to the same degree as the mammalian inactive X since the aim of this form of dosage compensation is to merely reduce the level of transcription of X-linked genes not to silence them completely. Perhaps the resultant chromatin structure would be more akin to G-band euchromatin than heterochromatin?

Drosophila utilizes yet another mechanism for ensuring dosage compensation—the single X of males is overactive compared to the two Xs of females and to the autosomes. Cytologically, polytenized hyperactive Xs look more diffuse and bloated than the other chromosomes, suggestive of a less compact and more accessible chromatin structure.[29] Perhaps we should think about this chromatin as at the opposite extreme of flavors to that of heterochromatin, and the hyperactive X as a sort of extreme

T-band. We will again consider the general and specialized chromosomal proteins involved in this process in subsequent sections.

It is considered that a form of facultative heterochromatin in normal *Drosophila* development is the silencing of the homeotic loci (e.g., the bithorax complex) that are involved in setting up the body plan of the insect. This silencing occurs along the anterior-posterior axis of the developing embryo and some of the *Polycomb* group (*Pc*-G) gene products that effect this silencing have domains (e.g., the chromo-domain) in common with other heterochromatin proteins. The sites of silenced homeotic loci on polytene chromosomes are characterized by constriction and slight under-replication[30] and at a molecular level the resulting chromatin structure can be considered as facultative heterochromatin.

Heterochromatin can spread and impose itself on flanking regions of euchromatin inactivating nearby genes. The resulting overt phenotype—position effect variegation (PEV) reflects the balance between the heritability of the inactive heterochromatin state and the ability of this to switch to another metastable chromatin state that is permissive for transcription. We have classed this phenomenon as facultative heterochromatin because of the ability of the silent state to switch to an active one, though this process appears to be stochastic and cannot be claimed to be under developmental control unlike the other forms of facultative heterochromatin that we have considered. PEV can be experimentally induced and the factors controlling it dissected genetically. This has been done to most advantage in the yeasts *S. cerevisiae* and *S. pombe* and in *Drosophila*. The elegance of these genetic systems has illustrated the role that both general and specialized chromosomal proteins play in the formation of heterochromatin. In *Drosophila,* where at least some heterochromatin can be seen on chromosomes, euchromatic polytene chromosome bands that have been juxtaposed next to constitutive heterochromatin are seen to disappear and merge into the block of heterochromatin. In some, but not all instances, this is accompanied by underpolytenization of the 'euchromatic' sequences.[31-33] There is no evidence for the spreading of C-band material into euchromatin in mammals but this may just be a problem of cytogenetic resolution.

GENERAL CHROMOSOMAL PROTEINS AND HETEROCHROMATIN

Constitutive heterochromatin is the site of most methylated cytosine in human and mouse chromosomes[34] and so the ^{5Me}CpG binding protein MECP2 is highly concentrated over constitutive heterochromatin but not over facultative heterochromatin in these vertebrates.[35] However, the methylation pattern of facultative heterochromatin can be distinctive. In placental mammals, but not marsupials the CpG islands associated with genes subject to inactivation are methylated on the inactive X[36] and in some cases the inactivation, late replication and condensation of this chromosome can be reversed by inhibition of DNA methylation with 5-azacytidine.[37,38] Genes on the inactive X reactivated in this manner demethylate their CpG islands.[36]

Moving up from DNA modification to the first level of chromatin structure, a common feature of constitutive heterochromatin is that the histones within it are, in general underacetylated.[39] Exceptions to this are in *Drosophila* polytene chromosomes where the heterochromatin at the chromocenter is enriched in histones acetylated at one particular residue—lysine 12,[40] and in mammalian embryonic stem cells where, prior to differentiation, histone H4 within the pericentromeric heterochromatin is acetylated. After differentiation this heterochromatin becomes hypoacetylated as in other cell types.[41] Nevertheless, the centromeric heterochromatin in undifferentiated ES cells still looks just like heterochromatin cytologically, i.e., it is condensed throughout the cell cycle and C-bands at metaphase so deacetylated histones are not a prerequisite for heterochromatin formation per se. It is also not clear how the histone acetyl transferase and deacetylase enzymes that have been identified so far can be targeted into this heterochromatin (chapter 4). Since these changes in heterochromatin are under developmental control should we call this heterochromatin constitutive or facultative?

Typical facultative heterochromatin is also associated with histone H2A, H3 and H4 hypoacetylation.[39] The inactive mammalian X has underacetylated histone H4 except at a few bands that may contain genes that escape X inactivation.[42] Nothing is known about the role of histone acetylation in the *C. elegans* dosage compensation mechanism, however, nucleosomes on the hyperactive X (what we have considered to be the opposite end of the spectrum from heterochromatin) in male *Drosophila* are enriched in histone H4 acetylated at lysine 16.[40]

Genetic and environmental modifiers of heterochromatin-mediated gene silencing in both *Drosophila* and *S.cerevisiae* result in increased levels of histone acetylation, suggesting that the silent chromatin is characterized by underacetylated histones.[43-45] In addition, nuclease accessibility studies indicate that such silenced regions may be packaged into more regular arrays of nucleosomes, or nucleosomal arrays that are less accessible to enzymes that act on the underlying DNA than is found at most euchromatic sites.[46-48] This may not be universally true of facultative heterochromatin since there are no reported differences in chromatin accessibility at homeotic loci that have been silenced by *Pc*-G products.[49]

Direct genetic evidence also implicates histones as a key component in the formation of heterochromatin. Deletion of some of the copies of the histone genes decreases heterochromatin-mediated repression.[50] Sites in yeast (both *S. cerevisiae* and *S. pombe*) at which such repression is monitored are those that occur naturally (the silent mating type loci) and those that have been manipulated experimentally (centromeres and telomeres). Acetylatable lysines in the N-terminus of histone H4 are needed for silencing at the mating type loci.[51] Revealingly, mutations in similar residues in H3 have a lesser effect on silencing at the mating type loci though they do affect silencing at telomeres, and they exert their main phenotype on the inducibility of genes located in nonheterochromatic parts of the yeast genome. The N terminus of H3 cannot substitute for that of H4 in the restoration of silencing.[52]

Clearly the nucleosomal structure of heterochromatin, both constitutive and facultative, is distinctive. It is less clear to what extent, if any, heterochromatin impinges on the higher orders of chromatin packaging or vice versa. There is no obvious difference in the association of constitutive heterochromatin with the core of extracted mitotic chromosomes as compared with that of euchromatin, except at a specialized heterochromatic site—the centromere.[53] Additionally, no differences in scaffold association could be found for sequences on the inactive or active Xs in mammals.[54]

SPECIALIZED PROTEINS OF HETEROCHROMATIN

While the genetic analysis of heterochromatin has been useful for illustrating the role of general chromosomal proteins in the inactive state, the real power of this approach has been its ability to identify new proteins that seem to be specific for heterochromatin. Some of these proteins are now known to interact directly with histones while the mechanism of action of the others remains unknown.

Although the inactive X of mammals is associated with a specific RNA molecule[27] and has hypoacetylated nucleosomes[39,42] no chromosomal protein has yet been identified that binds specifically to this chromosome. It is entirely possible that no such protein exists and that the heterochromatic structure and behavior of the inactive X is initiated solely by XIST RNA and propagated through changes in chromatin structure at the level of general chromosomal proteins such as nucleosomes.

The one instance in which a putative protein of the chromosome scaffold has been implicated in the formation of a type of facultative heterochromatin, and in gene silencing, is in the dosage compensation pathway of *C. elegans*. One of the proteins important in this process is DPY-27—an SMC protein of the XCAP-C/cut3 family (see chapter 4). Unlike the other SMC proteins, DPY-27 specifically recognizes and binds to the X chromosomes and although we don't know the details of how this specificity is achieved it is dependent on the products of the sdc-2, sdc-3 and dpy-30 genes.[55] Sdc-3 is a zinc-finger protein that also shows X-specific localization.[56] It seems reasonable that DPY-27 contributes to dosage compensation by binding to, and increasing the condensation of, these X chromosomes in interphase cells and hence decreasing their transcription.[55] At this stage we do not know whether DPY-27 behaves as a chromosome scaffold protein biochemically.

The formation of the hyperactive, hyperacetylated male X chromosome of *Drosophila* is dependent on *male specific lethal* (*msl*) and *maleless* (*mle*) genes that are themselves controlled by the *sex-lethal* (*sxl*) gene. The Mle protein is probably an RNA helicase, Msl-2 contains a RING finger motif that might direct chromatin association. Mle and the each of the Msl proteins associates, probably in a heteromeric complex, exclusively with the hyperactive X to give a banded pattern on polytene chromosomes by immunofluorescence.[57] This pattern is very similar to that of hyperacetylated H4 molecules.[58] Both the hyperacetylation of H4 on the hyperactive X and the bloated cytological appearance of this chromosome are entirely dependent on the presence of Mle and the Msl proteins.

The maintenance of a repressed chromatin structure over homeotic genes is dependent on the products of the *Pc*-G genes. The eponymous protein polycomb (Pc) contains an amino acid motif—the chromo domain—that is also found in proteins of constitutive heterochromatin and that is necessary for Pc binding to the homeotic loci.[59,60] Two other *Pc*-G genes encode proteins polyhomeotic (Ph) and posterior sex combs (Psc) with zinc-finger-like motifs. Pc, Ph and polycomblike (Pcl) have identical binding patterns on polytene chromosomes,[61,62] whereas binding sites for Psc and another Pc-G protein suppressor 2 of zeste [Su(z)2] are overlapping but not identical.[63] Immunoprecipitation studies suggest that Pc is associated with many proteins in a very large complex that exerts its function at the level of higher-order chromosome structure and that has multiple contacts to the DNA sequence of the bithorax complex over the repressed domain.[61,64] The Pc complex is also thought to remain associated with mitotic chromosomes at these sites.[59] On polytene chromosomes the bithorax complex appears constricted and densely stained only in tissues in which it is silenced. Experimental disruption of the Pc complex results in locally decondensed polytene chromosomes at the expected sites.[63]

How are the repressive protein complexes targeted to the correct chromosomal loci? It has been proposed that two types of DNA element are needed for silencing: one for its initiation (SIL-I) and another type for its maintenance (SIL-M).[65] The segmentation gap gene products are thought to be the repressor proteins that bind to SIL-I elements and single out the repressed loci for long-term silencing. The Pc-G complex may then be targeted to the correct sites by the product of the *extra sex combs* (*esc*) *Pc*-G gene interacting with segmentation gap gene products directly or, more indirectly, with the chromatin structure of the repressed homeotic loci.[66] It is the SIL-M elements that may be important in this. The polycomb response element (PRE) is a sequence that seems to combine both the properties of SIL-I and SIL-Ms and it can direct the binding of Pc-G proteins to ectopic sites and silence neighboring genes.[67] Remembering that the facultative heterochromatin of the mammalian inactive X chromosome is coated by XIST RNA it is tantalizing that an RNA molecule may also be a component of the Pc-G complex.[68]

There is an antipodal group of genes—the trithorax group (*Trx*-G) that are involved in the maintenance of an active state at the homeotic loci. One member of this group—*brahma*—is a DNA-dependent ATPase. Its homology to the yeast SWI2/SNF2 facilitator of transcriptional activation suggests a proactive role in altering chromatin structure and that it may be capable of dismantling heterochromatin.[69] Another *Trx*-G member *trithorax-like* (*trl*) is the GAGA transcriptional activator,[70] yet it also localizes to pericentric chromatin and is important for mitotic chromosome function.[71] Trx-G and Pc-G complexes share some binding sites on polytene chromosomes and, in fact, trx binding to chromosomes is dependent on E(z)—one of the Pc-G proteins. Hence, there seems to be quite a complex relationship between the maintenance and disruption of the silenced state at some loci.[72]

That the repressed state has some analogies with that of constitutive heterochromatin is evidenced by the observation that some *Pc*-G and *Trx*-G genes are modifiers of PEV.[70,73] Similar screens have identified other specialized heterochromatin proteins mutations in which either enhance E(var) or suppress Su(var) the formation of the heterochromatin complex.[74] These proteins are diverse and include those that may have DNA binding domains, e.g., Su(var) 3-7 and modulo,[75,76] protein phosphatases, e.g., Su(var) 3-6,[77] and proteins required for DNA replication, e.g., mus209/PCNA.[78] However, the most interesting class of protein identified in this way has been chromo-domain containing proteins, e.g., Su(var) 2-5/HP1[78] and Su(var) 3-9 that contains a domain found in trx as well as a chromo domain.[79] These proteins clearly establish a link between the developmentally programmed silencing of homeotic loci, and the experimentally induced heterochromatin of PEV.

The dosage-sensitivity of the modifiers of PEV has led to the suggestion that, like *Pc*-G genes, they encode subunits of macromolecular complexes or that in both cases the proteins bind to different sites along the DNA (e.g., PREs) and then associate with each other to loop up the DNA and compact it.[80] Unlike the *Pc*-G products, both GAGA and HP1 can be seen localized to the chromocenter of *Drosophila* polytene chromosomes.[70,81] A chimeric HP1 protein in which the endogenous chromo-domain has been replaced by that from Pc binds to both the chromocenter and to Pc binding sites on the chromosome arms. Moreover, endogenous Pc becomes mis-localized to pericentric heterochromatin,[60] a demonstration that heterochromatin protein localization is not simply determined by DNA sequence alone (there are no PREs in the chromocenter) but is influenced by protein-protein interactions.

Several HP1-like chromo-domain containing proteins have been identified in mammals[82] and they localize to sites of constitutive heterochromatin in interphase nuclei and metaphase chromosomes in a pattern coincident with that of C-bands.[83] Mutations in *bmi-1,* a murine homologue of *Psc,* result in a homeotic transformation phenotype,[84] and some genes inserted into pericentric heterochromatin in mammals can be shown to undergo PEV.[85] Hence, it is likely that multimeric complexes of heterochromatin proteins are important in gene silencing and chromosome structure of vertebrates.

Transcriptional silencing at centromeres and silent mating type loci in the yeast *S. pombe* also requires a chromo-domain containing protein—Swi6—and so may operate via the formation of repressive heterochromatin complexes like those in *Drosophila*.[86] However, silencing phenomena studied in the yeast *S. cerevisiae* appear to involve a completely different suite of proteins that are unrelated to those we have just discussed—the reason for this is unclear at present. Aside from mutations within the histone genes,[51,52] mutations in *RAP1*, *SIR1*, *SIR2*, *SIR3* and *SIR4* disrupt silencing in *S. cerevisiae*. Silencer DNA elements that are necessary in cis for silencing at some sites in the genome have been defined and they contain binding sites for Rap1p, Abf1p and the origin-recognition complex ORC (chapter 5). Sir1p and the proteins binding to the silencers are considered necessary to set up silencing but are not thought to be integral to heterochromatin itself,[87] whereas Sir2p, Sir3p and Sir4p are considered to be essential components of the heterochromatic complex itself. The role of ORC seems to be to target Sir1p to the silencer. It has been suggested that this is the initiating event in the formation of heterochromatin and leads to the recruitment of Sir3p and Sir4p.[88] Sir3p and Sir4p appear to interact directly with the N termini of histones H3 and H4.[89] Rap1p, that paradoxically is also a transcriptional activator, may interact directly with Sir3p and Sir4p circumventing the need for Sir1p at some sites in the genome.[90]

There are a bewildering array of proteins with roles in the formation of heterochromatin. Does this reflect the special molecular complexity of the heterochromatic state as compared with that of euchromatin or is it merely ascertainment bias reflecting the power of the genetic screens that can be used to identify protein components of heterochromatin? If the latter is the case will there be an equally large number of specialist chromosomal proteins found in euchromatin? Our growing knowledge about the very large multimeric complexes involved in transcription and transcriptional regulation of euchromatic genes (reviewed in refs. 91, 92) suggests that the answer to this question is yes.

CENTROMERES AND TELOMERES

So far we have only considered a role for heterochromatin in the stable transcriptional repression of genetic loci. However, there is accumulating evidence that the special chromatin structures found in heterochromatin may have important roles at centromeres and telomeres in maintaining the stability and integrity of the chromosome.

The chromatin structure of normal mammalian centromeres is manifest cytologically not only as C-banded constitutive heterochromatin but also as the most striking morphological feature of the metaphase chromosome—the primary constriction. Are the C-banding, the presence of heterochromatin and the primary constriction linked to each other and/or to centromere function?

The kinetochore, where spindle microtubules are captured by the chromosome, coats the centromeric chromatin—most DNA is under the inner kinetochore plate.[93] In some mammals a protein—CENP-B—can specifically bind a sequence embedded within the repeating unit of human alphoid and mouse minor satellite arrays in the centromeric DNA. However, CENP-B can bind alphoid DNA inserted at other chromosomal sites that, although they may in some cases adopt a constricted appearance, do not function as centromeres.[94,95] Moreover, inactive centromeres may still C-band but may not form constrictions.[96] On the other hand, some apparently completely functional human centromeres are C-band negative and contain no detectable alphoid DNA or CENP-B protein, but they do form a constriction.[97] Therefore the presence of a constriction seems to be necessary, but not sufficient for centromere formation in man.

Another kinetochore protein CENP-A appears to be a centromere-specific form of histone H3. In addition to the hypoacetylation of histone H4 in pericentric heterochromatin,[41] this suggests that the chromatin structure of centromeres may be distinctive even at the level of the components of the nucleosome.[98]

The association of centromeric heterochromatin with the chromosome core appears to be unusual and may reflect the biochemical fractionation of many kinetochore proteins with this mitotic chromosome scaffold (chapter 4).[53] It is this higher order organization of the centromere that may contribute to its constricted appearance on metaphase chromosomes. CENP-C and CENP-E are components of the mitotic scaffold and of the inner and outer kinetochore plate, respectively. They are specific for functional centromeres and so may contribute to the unique morphology of active centromeres.[95]

Nuclease studies have shown that the chromatin conformation of centromeric DNA is unusual in both budding and fission yeasts and in the latter case it can repress the transcription of genes placed there.[99-101] Furthermore, it seems that the presence of heterochromatin at the centromere is a critical component of a functional centromere.[102]

Telomeres are also characterized by unusual chromatin configuration, although there are no chromosome banding techniques that highlight them specifically. The ends of *S. cerevisiae* telomeres are devoid of nucleosomes and bound by a specialized protein complex—the teleosome—with the adjacent DNA being packaged into hypoacetylated nucleosomes.[103] The very ends of telomeres of other eukaryotes are similarly organized in non-nucleosomal nuclease resistant chromatin and specialized proteins have been found at the telomere ends.[104,105] More internally, nuclease analysis of subtelomeric DNA suggests the presence of nucleosomes.[104] Heterochromatin is also present at the telomeres of at least some eukaryotic chromosomes and can elicit PEV on genes located nearby.[106-108] However, in contrast to centromeres, it is unclear whether the formation of heterochromatin is actually of any importance for telomere function per se. One possible function of heterochromatin at telomeres might be to hide the ends of DNA molecules from cell-cycle check point mechanisms that monitor for the presence of double-strand breaks in the genome.

There are considerable similarities between PEV induced by telomeres and other sites in the genome,[109] and chromo domain proteins can be visualized there.[102] It has even been suggested that physical proximity to telomeres may be influencing repression at the silent mating type loci in yeast.[51] However, some proteins needed for silencing at other sites in the genome are dispensable for silencing at telomeres.[109] Different sites of heterochromatin may not be equivalent with each other, instead they might be considered to have different flavors just as euchromatin does. PEV at pericentric or fourth chromosome sites in *Drosophila* can be distinguished genetically from that at telomeres,[46] and telomeric heterochromatin does not C-band.[110] We might therefore consider telomeric heterochromatin as a rather weak form of heterochromatin.

The cytogenetic behavior of telomeres is unusual at meiosis. Most human telomeres are located at the ends of mitotic T-bands. Interstitial T- and R-bands correspond to the interchromomeric regions of meiotic chromosomes, it is the G-bands that correspond with chromomeres. However, telomeres appear to be in chromomeres at meiosis[111] and this may result from specialized functions of the telomere at meiosis.

In some unusual situations telomeres appear to be able to take on the job of the centromere. In the mitotic divisions of *C. elegans* spindle microtubules do not attach to the chromosomes at a discrete centromere but to the entire length of the (holocentric) chromosome. However, in meiotic divisions spindle attachments seem to be made to one of the pair of telomeres on each chromosome.[112]

HETEROCHROMATIN AND CELL CYCLE EVENTS

The classical cytological definition of heterochromatin is that it remains condensed throughout the cell cycle unlike euchromatin. There are other facets of heterochromatin biology during the cell cycle that also distinguish it from euchromatin.

Late replication is often considered to be an almost universal feature of heterochromatin. Mammalian heterochromatin is generally thought to be late replicating though there is considerable dispute about the time of replication of mammalian centromeric DNA.[113,114] Some of these discrepancies might arise through access problems to this chromatin with anti-BrdU antibodies. Human telomeres can replicate early or late—in the latter case even if cytogenetically they are located in T-bands (chapter 5).[115] In *S. cerevisiae*, silent mating type loci and telomeric sequences (sites of heterochromatin mediated gene silencing) replicate late and the latter remain late even when located internally, whereas centromeres replicate early (and in this organism centromeres do not repress the transcription of genes located within them and so may not be classed as heterochromatic).[116] In *Drosophila*, homeotic loci replicate late in salivary glands where they are silenced, but not in expressing tissues.[117] Loci subject to PEV and the sequences of α- (but not β-) heterochromatin are actually under-replicated compared to the rest of the genome in polytenized chromosomes,[118] and a component of the replication machinery is a modifier of PEV.[78] Polytene under-replication may result from an actual inhibition of replication of heterochromatin rather than just a reflection of delayed replication during the polytenization process.[119] The most direct biochemical link between heterochromatinization and DNA replication is the pivotal role of the ORC in silencing in *S. cerevisiae*,[88] though the replication and silencing functions of ORC are separable genetically.[120] Similar to the role of *esc* in silencing of homeotic loci in the fly, heterochromatinization in yeast can be initiated by recruitment of Sir1p by the ORC: Sir1p is then dispensable for maintenance of the silenced state once established[88] but it is needed for the inheritance of the silenced state in cycling cells. These experiments indicate that heterochromatin is not self-templating,[87] and analysis of telomeric silencing in yeast indicate that S-phase and metaphase may be stages of the cycle at which heterochromatin is vulnerable to dismantling.[121] Restoring repression at the yeast silent mating type loci requires at least one passage through S phase,[122] suggesting that this stage of the cycle also provides a window of opportunity for heterochromatin formation. ORC is required during metaphase to maintain the silenced state,[123] hence the distinctive appearance of heterochromatin on mitotic chromosomes should not merely be considered a passive consequence of the interphase state but be viewed as an active component in the maintenance of a heritable repressed chromatin state. Late replication may play an active role in propagation of heterochromatin by directing this chromatin into specialized factories or parts of the nucleus.[114]

Heterochromatin is usually considered to be adjacent to the nuclear membrane, though conflicting reports indicate that this may depend on cell cycle stage and cell type (chapter 1). Rap1p in *S. cerevisiae* is associated with telomeres in nuclei and on meiotic chromosomes,[124] and in nuclei it is located preferentially at the nuclear periphery in a pattern indicative of telomere clustering that is thought to be mediated though interaction with Sir3p and Sir4p.[125] In *Drosophila*, the nuclear position of genes which have had heterochromatin-forming satellite inserted into them and that are subject to PEV, as well as their unaltered homologues, become mis-localized to sites of heterochromatin in the nucleus.[126] Detailed analysis shows that the localization is in fact to centromeric heterochromatin from the same chromosome and that it is not just a generalized heterochromatin association.[127] This localization is dependent on Su(var)s and E(var)s and cell-cycle and developmental stage[126,127] suggesting that heterochromatin and its constituent proteins play a very precise and proactive role in the organization of the interphase nucleus.

Although we have considered constitutive heterochromatin as essentially always present in contrast to facultative heterochromatin, it is in fact under developmental control. No condensed heterochromatin is visible in early embryonic development of *Drosophila* or amphibians. In *Drosophila* early embryos the GAGA factor, that directly binds a distinct DNA satellite sequence, is localized to restricted sites in the nucleus at what are assumed to be sites of presumptive heterochromatin. By way of contrast HP1, that is thought to bind to heterochromatin through protein-protein interactions, shows no such localization at this stage.[128] The recruitment of HP1 to a conspicuous chromocenter at cycle 14 is accompanied by increased phosphorylation of the protein.[129] GAGA factor in flies is also clearly present at the chromocenter of mitotic chromosomes whereas, unlike the localization of chromo-domain proteins on mammalian chromosomes,[83] HP1 seems to be dispersed from the chromosome at metaphase.[128] However, HP1 is needed during prophase for correct chromosome condensation and ensuing segregation.[130] Analysis of the silenced state through development suggests a positive role for mitosis in the formation of heterochromatin and the silenced state.[131] Silencing (and by inference heterochromatin) was found to be mitotically stable, whereas the cessation of cell division in terminally differentiating tissues resulted in a relief from silencing—indeed this was suggested to explain some forms of variegation.

The last characteristic of heterochromatin that we shall consider is its behavior at meiosis. In mammals, heterochromatin occupies a smaller fractional length of pachytene chromosomes than it does of mitotic metaphase chromosomes and the synaptonemal complex at sites of heterochromatin looks different by EM, lacking recombination nodules.[132] There is evidence that recombination is suppressed by at least 10-fold across the centromeric heterochromatin of mammalian chromosomes.[133] Decreased recombination is also a feature of silenced sites in yeast heterochromatin and this is relieved by mutations in the very same proteins that are involved in transcriptional silencing.[86,134] It is assumed that decreased recombination is a direct reflection of the inaccessibility of heterochromatin and that the absence of recombination contributes to the accumulation of repetitive DNA at sites of heterochromatin. This brings us full circle because we argued earlier that arrays of tandem repeats may have a propensity to form heterochromatin.

REFERENCES

1. Heitz E. Das Heterochromatin der Moose. I. Jahrb. Wiss. Bot. 1928; 69:762-818.
2. Alfi OS, Donnell GN and Derencsenyi A. C-banding of human chromosomes produced by DNase. Lancet 1973; 1:505.
3. Burkholder GD. Morphological and biochemical effects of endonucleases on isolated mammalian chromosomes in vitro. Chromosoma 1989; 97:347-355.
4. Dorer DR and Henikoff S. Expansions of transgene repeats cause heterochromatin formation and gene silencing in Drosophila. Cell 1994; 77: 993-1002.
5. Simon D and Knowles BB. Newly acquired peri-telomeric heterochromatin in a transgenic mouse lineage. Cytogenet. Cell Genet. 1993; 62:211-213.
6. Kunze B, Weichenham D, Virks P et al. Copy numbers of a clustered long-range repeat determine C-band staining. Cytogenet. Cell Genet. 1996; 73:86-91.
7. Kovacs G. Homogeneously staining regions on marker chromosomes in malignancy. Int. J. Cancer 1979; 23:299-301.
8. Burkholder GD, Latimer LJP and Lee JS. Immunofluorescent staining of mammalian nuclei and chromosomes with a monoclonal antibody to triplex DNA. Chromosoma 1988; 97:185-192.
9. Miklos GLG and Cotsell JN. Chromosome structure at interfaces between major chromatin types: alpha and beta heterochromatin. BioEssays 1990; 13: 1-6.
10. Pimpinelli S, Berloco M, Fanti L et al. Transposable elements are stable structural elements of Drosophila melanogaster heterochromatin. Proc. Natl. Acad. Sci. USA 1995; 92:3804-3808.
11. Lohe AR and Hilliker AJ. Return of the H-word (heterochromatin). Curr. Opin. Genet. Develop. 1995; 5:746-755.
12. Craig JM and Bickmore WA. The distribution of CpG islands in mammalian chromosomes. Nature Genet. 1994; 7:376-382.
13. Wakimoto BT and Hearn MG. The effects of chromosome rearrangements on the expression of heterochromatic genes in chromosome 2L of *Drosophila melanogaster*. Genet. 1990; 125:141-151.
14. Eberl DF, Duyf BJ and Hilliker AJ. The role of heterochromatin in the expression of a heterochromatic gene, the *rolled* locus of *Drosophila melanogaster*. Genet. 1993; 134:227-292.
15. Gatti M and Pimpinelli S. Functional elements in *Drosophila*

melanogaster heterochromatin. Ann. Rev. Genet. 1992; 26:239-275.
16. Crouse HV. The controlling element in sex chromosome behavior in *Sciara*. Genet. 1960; 45:1425-1443.
17. Brown SW and Nur U. Heterochromatic chromosomes in the coccids. Science 1964; 145: 130-136.
18. Nakai Y and Kohno S. Elimination of the largest chromosome pair during differentiation into somatic cells in the Japanese hagfish, Myxine garmani. Cytogenet. Cell Genet. 1987; 45: 80-83.
19. Kanda N. A new differential technique for staining the heteropycnotic X chromosome in female mice. Exp. Cell Res. 1973; 80:463-467.
20. Sarto GE, Therman E and Patau K. Increased Q fluorescence of an inactive Xq- chromosome in man. Clin. Genet. 1974; 6:289-293.
21. Yoshida I, Kashio N and Takagi N. Cell fusion-induced quick change in replication time of the inactive mouse X chromosome: an implication for the maintenance mechanism of late replication. EMBO J. 1993; 12:4397-4405.
22. Camargo M and Cervenka J. Patterns of DNA replication of human chromosomes. II. Replication map and replication model. Am. J. Hum. Genet. 1982; 34:757-780.
23. Drouin R, Lemieux N and Richer C-L. Analysis of DNA replication during S phase by means of dynamic chromosome banding at high resolution. Chromosoma 1990; 99: 273-280.
24. Walker CL, Cargile CB, Floy KM et al. The Barr body is a looped X chromosome formed by telomere association. Proc. Natl. Acad. Sci. USA 1991; 88:6191-6195.
25. Rastan S. Non-random X-chromosome inactivation in mouse X-autosome translocation embryos: location of the inactivation centre. J. Embryol. Exp. Morphol. 1983; 78:1-22.
26. Penny GD, Kay GF, Sheardown SA et al. Requirement for *Xist* in X chromosome inactivation. Nature 1996; 379:131-137.
27. Brown CJ, Hendrich BD, Rupert JL et al. The human XIST gene: analysis of a 17 kb inactive X-specific RNA that contains conserved repeats and is highly localized within the nucleus. Cell 1992; 71:527-542.
28. Brown CJ and Willard HF. The human X-inactivation centre is not required for maintenance of X-chromosome inactivation. Nature 1994; 368:154-156.
29. Dobzhansky T. The X-chromosome in the larval salivary glands of hybrids *Drosophila insularis* x *Drosophila tropicalis*. Chromosoma 1957; 8:691-698.
30. Spierer A and Spierer P. Similar level of polyteny in bands and interbands of *Drosophila* giant chromosomes. Nature 1984; 307:176-178.

31. Henikoff S. Position-effect variegation and chromosome structure of a heat shock puff in *Drosophila.* Chromosoma 1981; 83:381-393.
32. Zhimulev IF, Belyaeva ES, Fomina OV et al. Cytogenetic and molecular aspects of position-effect variegation in *Drosophila melanogaster.* Chromosoma 1986; 94:492-504.
33. Karpen G and Spradling A. Reduced DNA polytenization of a minichromosome region undergoing position-effect variegation in *Drosophila melanogaster.* Cell 1990; 63:97-107.
34. Miller OJ, Schnedl W, Allen J and Erlanger BF. 5-Methylcytosine localized in mammalian constitutive heterochromatin. Nature 1974; 251:636-637.
35. Lewis JD, Meehan RR, Henzel WJ et al. Purification, sequence and cellular localization of a novel chromosome protein that binds to methylated DNA. Cell 1992; 69:905-914.
36. Toniolo D, Martini G, Migeon BR and Dono R. Expression of the G6PD locus on the human X chromosome is associated with demethylation of three CpG clusters within 100 kb of DNA. EMBO J. 1988; 7:401-406.
37. Haaf T, Ott G and Schmid M. Inhibition of condensation in the late-replicating X chromosome induced by 5-azadeoxycytidine in human lymphocyte cultures. Hum. Genet. 1988; 79:18-23.
38. Jablonkla E, Goiten R, Marcus M and Cedar H. DNA hypomethylation causes an increase in DNase-I sensitivity and an advance in the time of replication of the entire inactive X chromosome. Chromosoma 1985; 93:152-156.
39. Belyaev ND, Keohane AM and Turner BM. Differential underacetylation of histones H2A, H3 and H4 on the inactive X chromosome in human female cells. Hum. Genet. 1996; 97:573-578.
40. Turner BM, Birley AJ and Lavender J. Histone H4 isoforms acetylated at specific lysine residues define individual chromosomes and chromatin domains in *Drosophila* polytene nuclei. Cell 1992; 69:375-384.
41. O'Neill LP and Turner BM. Histone acetylation distinguishes coding regions of the human genome from heterochromatin in a differentiation-dependent but transcription-independent manner. EMBO J. 1995; 14:3946-3957.
42. Jeppesen P and Turner BM. The inactive X chromosome in female mammals is distinguished by a lack of histone H4 acetylation, a cytogenetic marker for gene expression. Cell 1993; 74:281-289.
43. Mottus R, Reeves R and Grigliatti TA. Butyrate suppression of position-effect variegation in *Drosophila melanogaster.* Mol. Gen. Genet. 1980; 178:465-469.
44. Dorn R, Heyman S, Lindigkeit R and Reuter G. Suppressor mutation of position effect variegation in *Drosophila melanogaster* affect-

ing chromatin properties. Chromosoma 1986; 93:398-403.
45. Braunstein M, Rose AB, Holmes SG, Allis CD and Broach JR. Transcriptional silencing in yeast is associated with reduced nucleosome acetylation. Genes Develop. 1993; 7:592-604.
46. Wallrath LL and Elgin SCR. Position effect variegation in *Drosophila* is associated with an altered chromatin structure. Genes Develop. 1995; 9:1263-1277.
47. Nasmyth KA. The regulation of yeast mating-type chromatin structure by SIR: An action at a distance affecting both transcription and transposition. Cell 1982; 30:567-578.
48. Gottschling DE. Telomere-proximal DNA in *Saccharomyces cerevisiae* is refractory to methyltransferase activity in vivo. Proc. Natl. Acad. Sci. USA 1992; 89:4062-4065.
49. Schlossherr J, Eggert H, Paro R, Cremer S and Jack RS. Gene activation in Drosophila mediated by the polycomb gene product or by position-effect variegation does not involve major changes in the accessibility of the chromatin fiber. Mol. Gen. Genet. 1994; 243:453-462.
50. Moore GD, Sinclair DA and Grigliatti TA. Histone gene multiplicity and position effect variegation in *Drosophila melanogaster*. Genet. 1983; 105:327-344.
51. Thompson JS, Ling X and Grunstein M. Histone H3 amino terminus is required for telomeric and silent mating locus repression in yeast. Nature 1994; 369; 245-247.
52. Ling X, Harkness TAA, Schultz MC et al. Yeast histone H3 and H4 amino termini are important for nucleosome assembly in vivo and in vitro: redundant and position-independent functions in assembly but not in gene regulation. Genes Develop. 1996; 10:686-699.
53. Bickmore WA and Oghene K. Visualizing the spatial relationship between defined DNA sequences and the axial region of extracted mitotic chromosomes. Cell 1996; 84:95-104.
54. Beggs AH and Migeon BR. Studies of chromatin loop structure of the human X chromosome: relevance to X inactivation and CpG clusters. Mol. Cell Biol. 1989; 9:2322-2331.
55. Chuang PJ, Albertson DG and Meyer BJ. DPY-27: A chromosome condensation protein homolog that regulates *C. elegans* dosage compensation through association with the X chromosome. Cell 1994; 79:459-474.
56. Ryner LC and Swain A. Sex in the 90s. Cell 1995; 81:483-493.
57. Kelley RL and Kuroda MI. Equality for X chromosomes. Science 1995; 270:1607-1610.
58. Bone JR, Lavender JS, Richman R. et al. Acetylated histone H4 on the male X chromosome is associated with dosage compensa-

tion in *Drosophila*. Genes Develop. 1994; 8:96-104.
59. Messmer S, Franke A and Paro R. Analysis of the functional role of the *polycomb* chromo domain in *Drosophila melanogaster*. Genes Develop. 1992; 6:1241-1254.
60. Platero JS, Hartnett T and Eissenberg JC. Functional analysis of the chromodomain of HP1. EMBO J. 1995; 14:3977-3986.
61. Francke A, DeCamillis M, Zink D et al. Polycomb and polyhomeotic are constituents of a multimeric protein complex in chromatin of *Drosophila melanogaster*. EMBO J. 1992; 11:2941-2950.
62. Lonie A, D'Andrea R, Paro R and Saint R. Molecular characterization of the *Polycomblike* gene of *Drosophila melanogaster* a trans-acting negative regulator of homeotic gene expression. Develop. 1994; 120:2629-2636.
63. Rastelli L, Chan CS and Pirotta V. Related chromosome binding sites for zeste, suppressors of zeste and Polycomb group proteins in Drosophila and their dependence on Enhancer of zeste function. EMBO J. 1993; 12:1513-1522.
64. Orlando V and Paro R. Mapping polycomb-repressed domains in the bithorax complex using in vivo-formaldehyde cross-linked chromatin. Cell 1993; 75:1187-1198.
65. Bienz M and Muller J. Transcriptional silencing of homeotic genes in Drosophila. BioEssays 1995; 17:775-784.
66. Simon J, Chang A, Bender W et al. Elements of the *Drosophila* bithorax complex that mediate repression by *Polycomb* group products. Dev. Biol. 1993; 158:131-144.
67. Zink D and Paro R. *Drosophila* Polycomb-group regulated chromatin inhibits the accessibility of a trans-activator to its target DNA. EMBO J. 1995; 14:5660-5671.
68. Hendrich BD and Willard HF. Epigenetic regulation of gene expression: the effect of altered chromatin structure from yeast to mammals. Hum. Mol. Genet. 1995; 4:1765-1777.
69. Tamkun JW, Deuring R, Scott MP et al. Brahma: a regulator of *Drosophila* homeotic genes structurally related to the yeast transcriptional activator SNF2/SWI2. Cell 1992; 68:561-572.
70. Farkas G, Gausz J, Galloni M et al. The *Trithorax-like* gene encodes the *Drosophila* GAGA factor. Nature 1994; 371;806-808.
71. Bhat KM, Gausz J, Farkas G et al. The GAGA factor is required in the early *Drosophila* embryo not only for transcriptional regulation but also for nuclear division. Develop. 1996; in press.
72. Kuzin B, Tilib S, Sedkov Y et al. The *Drosophila* trithorax gene encodes a chromosomal protein and directly regulates the region-specific homeotic gene Fork-head. Genes Develop. 1994; 8:2478-2490.
73. Grigliatti T. Position-effect variegation—an assay for nonhistone

chromosomal proteins and chromatin assembly and modifying factors. Methods Cell Biol. 1991; 35:587-627.
74. Karpen GH. Position-effect variegation and the new biology of heterochromatin. Curr. Opin. Genet. Develop. 1994; 4:281-291.
75. Reuter G, Giarre M, Farah J et al. Dependence of position-effect variegation in *Drosophila* on dose of a gene encoding an unusual zinc-finger protein. Nature 1990; 344: 219-223.
76. Garzino V, Pereira A, Laurenti P et al. Cell lineage-specific expression of *modulo*, a dose-dependent modifier of variegation in *Drosophila.* EMBO J. 1992; 11:4471-4479.
77. Drombardi V and Cohen PTW. Protein phosphorylation is involved in the regulation of chromatin condensation during interphase. FEBS Lett. 1992; 312: 21-26.
78. Henderson DS, Banga SS, Grigliatti TA and Boyd JB. Mutagen sensitivity and suppression of position-effect variegation result from mutations in *mus209*, the Drosophila gene encoding PCNA. EMBO J. 1994; 13:1450-1459.
79. Eissenberg JC, Morris GD, Reuter G and Hartnett T. The heterochromatin-associated protein HP-1 is an essential protein in *Drosophila* with dosage-dependent effects on position effect variegation. Genet. 1992; 131:345-352.
79a. Tschiersch B, Hofman A, Krauss V et al. The protein encoded by the *Drosophila* position-effect variegation suppressor gene *Su(var)3-9* combines domains of antagonistic regulators of homeotic gene complexes.
80. Henikoff S. Dosage-dependent modification of position-effect variegation in Drosophila. BioEssays 1996; 18:401-409.
81. Powers JA and Eissenberg JC. Overlapping domains of the heterochromatin-associated protein HP1 mediate nuclear localization and heterochromatin binding. J. Cell Biol. 1993; 120:291-299.
82. Saunders WS, Chue C, Goebl M. et al. Molecular cloning of a human homologue of Drosophila heterochromatin protein HP1 using anti-centromere autoantibodies with anti-chromo specificity J. Cell Sci. 1993; 104:573-582.
83. Wreggett KA, Hill F, James PS et al. A mammalian homologue of Drosophila heterochromatin protein (HP1) is a component of constitutive heterochromatin. Cytogenet. Cell Genet. 1994; 66:99-103.
84. Van der Lugt NMT, Domen J, Linders K et al. Posterior transformation, neurological abnormalities, and severe hematopoietic defects in mice with a targetted deletion of Bmi-1 proto-oncogene. Genes Develop. 1994; 8:757-769.
85. Festenstein R, Tolaini M, Corbella P et al. Locus control region function and heterochromatin-induced position effect variegation. Science 1996; 271:1123-1125.

86. Allshire RC, Nimmo ER, Ekwall K et al. Mutations derepressing silent centromeric domains in fission yeast disrupt chromosome segregation. Genes Develop. 1995; 9:218-233.
87. Holmes SG and Broach JR. Silencers are required for inheritance of the repressed state in yeast. Genes Develop. 1996; 10:1021-1032.
88. Triolo T and Sternglanz R. Role of interactions between the origin recognition complex and SIR1 in transcriptional silencing. Nature 1996; 381:251-253.
89. Hecht A, Laroche T, Bolsinger SS et al. Histone H3 and H4 N termini interact with SIR3 and SIR4 proteins: a molecular model for the formation of heterochromatin in yeast. Cell 1995; 80:583-592.
90. Moretti P, Freeman K, Coodly L and Shore D. Evidence that a complex of SIR proteins interacts with the silencer and telomere-binding protein RAP1. Genes Develop. 1994; 8:2257-2269.
91. Peterson CL, Dingwall A and Scott MP. SWI/SNF gene products are components of a large multisubunit complex required for transcriptional enhancement. Proc. Natl. Acad. Sci. USA 1994; 91:2905-2908.
92. Roth SY. Chromatin-mediated transcriptional repression in yeast. Curr. Opinions Genet. Develop. 1995; 5:168-173.
93. Cooke CA, Bazett-Jones DP, Earnshaw WC and Rattner JB. Mapping DNA within the mammalian kinetochore. J. Cell Biol. 1993; 120:1083-1091.
94. Haaf T, Warburton PE and Willard HF. Integration of human α-satellite DNA into simian chromosomes: centromere protein binding and disruption of normal chromosome segregation. Cell 1992; 70:681-696.
95. Sullivan BA and Schwartz S. Identification of centromeric antigens in dicentric Robertsonian translocations:CENP-C and CENP-E are necessary components of functional centromeres. Hum. Mol. Genet. 1995; 4:2189-2197.
96. Therman E, Trunca C, Kuhn EM and Sarto GE. Dicentric chromosomes and the inactivation of the centromere. Hum. Genet. 1986; 72:191-195.
97. Voullaire LE, Slater HR, Petrovic V and Choo KHA. A functional marker centromere with no detectable alpha-satellite, satellite III, or CENP-B protein: activation of a latent centromere? Am. J. Hum. Genet. 1993; 52:1153-1163.
98. Sullivan KF, Hechenberger M and Masri K. Human CENP-A contains a histone H3 related histone fold domain that is required for targetting to the centromere. J. Cell Biol. 1994; 127:581-592.
99. Heus JJ, Bloom KS, Zonneveld BJM. et al. Chromatin structures of *Kluyveromyces lactis* centromeres in *K. lactis* and *Saccharomyces*

cerevisiae. Chromosoma 1993; 102:660-667.

100. Marschall JG and Clarke L. A novel cis-acting centromeric DNA element affects *S. pombe* centromeric chromatin structure at a distance. J. Cell Biol. 1995; 128:445-454.
101. Allshire RC, Javerzat J-P, Redhead NJ and Cranston G. Position effect variegation at fission yeast centromeres. Cell 1994; 76:157-169.
102. Ekwall K, Javerzat J-P, Lorentz A et al. The chromdomain protein Swi6: a key component at fission yeast centromeres. Science 1995; 269:1429-1431.
103. Wright JH, Gottschling DE and Zakian VA. *Saccharomyces* telomeres assume a non-nucleosomal chromatin structure. Genes Develop. 1992; 6:197-210.
104. Tommerup H, Dousmanis A and de Lange T. Unusual chromatin in human telomeres. Mol. Cell Biol. 1994; 14:5777-5789.
105. Chong L, van-Steensel B, Broccoli D et al. A human telomeric protein. Science 1995; 270:1663-1667.
106. Levis R, Hazelrigg T and Rubin GM. Effects of genomic position on the expression of transduced copies of the *white* gene of *Drosophila.* Science 1985; 229:558-561.
107. Renauld H, Aparichio OM, Zierath PD et al. Silent domains are assembled continuously from the telomere and are defined by promoter distance and strength, and by SIR3 dosage. Genes Dev. 1993; 7:1133-1145.
108. Nimmo ER, Cranston G and Allshire RC. Telomere-associated chromosome breakage in fission yeast results in variegated expression of adjacent genes. EMBO J. 1994; 13:3801-3811.
109. Aparicio OM, Billington BL and Gottschling DE. Modifiers of position effect are shared between telomeric and silent mating-type loci in *S. cerevisiae.* Cell 1991; 66;1279-1287.
110. Pimpinelli S, Santini G and Gatti M. Characterization of Drosophila heterochromatin II. C- and N-banding. Chromosoma 1976; 57:377-386.
111. Ambros PF and Sumner AT. Correlation of pachytene chromomeres and metaphase bands of human chromosomes, and distinctive properties of telomeric regions. Cytogenet. Cell Genet. 1987; 44: 223-228.
112. Wicky C and Rose AM. The rôle of chromosome ends during meiosis in *Caenorhabditis elegans.* BioEssays 1996; 18:447-452.
113. Fox MH, Arndt-Jovin DJ, Jovin TM et al. Spatial and temporal distribution of DNA replication sites localized by immunofluorescence and confocal microscopy in mouse fibroblasts. J. Cell Sci. 1991; 99:247-253.
114. O'Keefe RT, Henderson C and Spector DL. Dynamic organization of DNA replication in mammalian cell nuclei: spatially and

temporally defined replication of chromosome-specific α-satellite sequences. J. Cell Biol. 1992; 116:1095-1110.
115. Ten Hagen KG, Gilbert DM, Willard HF and Cohen SN. Replication timing of DNA sequences associated with human centromeres and telomeres. Mol. Cell Biol. 1990; 10:6348-6355.
116. Friedman KL, Raghuraman MK, Fangman WL and Brewer BJ. Analysis of the temporal program of replication initiation in yeast chromosomes. J. Cell Sci. 1995; 19 (S):51-58.
117. Eissenberg JC, Elgin SCR and Paro R. Epigenetic regulation in *Drosophila*: a conspiracy of silence. In: Elgin SCR, ed. Chromatin Structure and Gene Expression. IRL Press, 1995:147-171.
118. Karpen GH and Spradling AC. Reduced DNA polytenization of a minichromosome region undergoing position-effect variegation in *Drosophila*. Cell 1990; 63:97-107.
119. Glaser RL, Karpen GH and Spradling AC. Replication forks are not found in a *Drosophila* minichromosome demonstrating a gradient of polytenization. Chromosoma 1992; 102:15-19.
120. Ehrenhofer-Murray AE, Gossen M, Pak DTS. et al. Separation of origin recognition complex functions by cross-species complementation. Science 1995; 270:1671-1674.
121. Aparicio OM and Gottschling DE. Overcoming telomeric silencing: a *trans*-activator competes to establish gene expression in a cell cycle-dependent way. Genes Develop. 1994; 8:1133-1146.
122. Miller AM and Nasmyth KA. Role of DNA replication in the repression of silent mating type loci in yeast. Nature 1984; 312:247-251.
123. Fox CA, Loo S, Dillin A and Rine J. The origin recognition complex has essential functions in transcriptional silencing and chromosomal replication. Genes Develop. 1995; 9:911-924.
124. Klein F, Laroche T, Cardenas ME et al. Localization of RAP1 and topoisomerase II in nuclei and meiotic chromosomes of yeast. J. Cell Biol. 1992; 117:935-948.
125. Palladino F, Laroche T, Gilson E et al. SIR3 and SIR4 proteins are required for the positioning and intergrity of yeast telomeres. Cell 1993; 75:543-555.
126. Csink AK and Henikoff S. Genetic modification of heterochromatic association and nuclear organization in *Drosophila*. Nature 1996; 381:529-531.
127. Dernberg AF, Broman KW, Fung JC et al. Perturbation of nuclear architecture by long-distance chromosome interactions. Cell 1996; 85:745-759.
128. Kellum R, Raff JW and Alberts BM. Heterochromatin protein 1 distribution during development and during the cell cycle in *Drosophila* embryos. J. Cell Sci. 1995; 108:1407-1418.

129. Eissenberg JC, Ge YW and Hartnett T. Increased phosphorylation of HP1, a heterochromatin-associated protein of *Drosophila,* is correlated with heterochromatin assembly. J. Biol. Chem. 1994; 269:21315-21321.
130. Kellum R and Alberts BM. Heterochromatin protein 1 is required for correct chromosome segregation in *Drosophila* embryos. J. Cell Sci. 1995; 108:1419-1431.
131. Lu BY, Bishop CP and Eissenberg JC. Developmental timing and tissue specificty of heterochromatin-mediated silencing. EMBO J. 1996; 15:1323-1332.
132. Stack SM. Heterochromatin, the synaptonemal complex and crossing over. J. Cell Sci. 1984; 71:159-176.
133. Wevrick R and Willard H. Long-range organization of tandem arrays of α-satellite DNA at the centromeres of human chromosomes: high frequency array length polymorphism and meiotic stability. Proc. Natl. Acad. Sci. USA 1989; 86:9394-9398.
134. Egel R, Willer M and Nielson O. Unblocking of meiotic crossing-over between the silent mating-type cassettes of fission yeast, conditioned by the recessive, pleiotropic mutant *rik1*. Curr. Genet. 1989; 15:407-410.

CHAPTER 7

The Chromosome Band: Origins, Perspectives and Future Prospects

We have looked at the techniques used to produce chromosome bands and presented a treatise of the structural and functional differences between different types, or "flavors" of chromosome bands. We have even suggested that regions of chromosomes from organisms with no conventional chromosome banding still fall into a number of distinctive chromatin flavors with familiar cosegregating properties. In this chapter we ask how and why these properties of DNA sequence, chromatin structure and chromosome behavior appear to have evolved together. Finally, we ask what the futures of both conventional chromosome banding and molecular cytogenetics are likely to be and how they will augment alternative techniques of genome analysis.

EVOLUTION OF CHROMOSOME BANDING

There is a spectrum of chromosome band types definable on either visible or molecular characteristics. These types of flavors extend from T-bands and R-bands through to G-bands and beyond to heterochromatin. The gross properties of these band types in mammals are summarized in Table 7.1.

Do all of the properties listed in Table 7.1 contribute actively to the final chromatin environment or are some of them just passive reporters of their surrounding environment and its behavior during the cell-cycle and over evolutionary time? If this is so, which of the features of Table 7.1 are the driving force that is dictating

Table 7.1. Major properties of mammalian chromosome bands

Property	T-bands	R-bands	G-bands	C-bands
Housekeeping genes	High density	High density	Few	None
Tissue-specific genes	High density	High density	Lower density	None
CpG islands	Very frequent	Quite frequent	Infrequent	Absent
SINES	Abundant copies of recently retroposed copies	Plentiful	Newly retroposed copies are infrequent	Infrequent
LINES	Relative paucity	Slight deficit	Abundant	Infrequent
Base composition	Most GC-rich	Relatively GC-rich	Relatively AT-rich	Mostly AT-rich but dependent on satellite families present
Histone H4 acetylation	Hyperacetylated and rapid turnover	Hyperacetylated	Levels of acetylation are low	Hypoacetylated
Topoisomerase II	Present at lower levels than in G-bands	Present at lower levels than in G-bands	Enriched	Concentrated in centromeric heterochromatin
Prophase to metaphase	Condenses and sub-bands fuse throughout metaphase	Condenses and fuses through metaphase	Condenses and sub-bands fused by prophase or early metaphase	Condensed at interphase condensation
Replication time	In first quarter of S	In first half of S	In second half of S	In second half of S
Meiotic recombination	High	High	Low	Suppressed
Appearance at meiosis I prophase	Interchromomeric regions (except for terminally located T-bands)	Inter-chromomeric regions	Chromomeres	Chromomeres

the trend in the other properties? To answer these questions we begin by turning to an evolutionary analysis of chromosome banding.

Table 7.2 summarizes whether standard chromosome banding techniques have been able to induce bands in the mitotic chromosomes from different groups of eukaryotes.[1] There are a few additional points worth noting. Firstly, although many mammals have been tested for banding patterns, the numbers of species tested within each group in Table 7.2 decreases down the table: the numbers of different bird, reptile, amphibia, fish and plant species tested is relatively small, generally less than ten in each case. In many cases discordant results are obtained among different species within the same group. Metaphase chromosomes from only one insect, *Drosophila*, have been studied so far, therefore these results should be extrapolated to other invertebrates with caution.

The interpretation of the patterns obtained with some chromosome banding methods also needs to be treated with care sometimes. G- or Q-bands in some species of amphibians, fish and plants may be identical to interspersed regions of heterochromatin and really correspond to C-bands rather than to standard euchromatic G-bands.[2,3]

The vast majority of mammalian and avian species examined to date have chromosomes that band easily with G-, and R-banding techniques.[1,4] In addition, most reptilian chromosomes band with G-, and R-banding techniques, although the quality of banding is variable.[1,5,6] With amphibia, fish and plants, some species band, others do not.[1-3,7-9] In general the lowest vertebrates with reported good G-banding are the bony fish.[10] It has been proposed that the chromosomes of some organisms may not band with some protocols because their chromatin is overcondensed.[11] It was originally proposed[1] that among fish, only eels produced euchromatic G-bands,[12] and that no plant chromosomes had euchromatic G-bands. It was subsequently shown that salmon show euchromatic banding patterns when digested with restriction enzymes[13] and plant euchromatin has been successfully banded by G-banding of prometaphase chromosomes[8] and by fluorescent G- and R-banding.[9] In addition, regions of differentially-condensing euchromatin have been observed in palm species.[14]

Table 7.2. Distribution of different types of banding patterns and presence of GC-rich isochores throughout eukaryotes

	Pachytene Chromomeres	Replication bands	Heterochromatic C-bands	G- or R-bands in Euchromatin	Euchromatic Q-bands	Isochores
mammals	+	+	+	+	+	+
birds	+	+	+	+	+	+
reptiles	+	+	+	+	+	+/-
amphibia	+	+	+	+	+/-	-
fish	+	+	+	+	+/-	+/-
insects*	nk	-	+	-	-	-
plants	+	+	+	-	+/-	+/-

+, chromosomes band/GC-rich isochores present in all species tested; +/-, chromosomes band/GC-rich isochores present in only some species tested; -, chromosomes do not band/GC-rich isochores not present in any species tested; *, only *Drosophila* studied so far; nk, not known. Data taken from refs. 1, 5, 8, 9, 12, 13, 14, 20.

All groups of eukaryotes that have chromosomes large enough to be visualized contain a proportion of heterochromatin, revealed by C-banding. The amount of heterochromatin seems to be independent of the genus type. For example, within the same frog genus, *Pleurodema*, one species *P. brachiops* has a genome size three and a half times larger than another species *P. thaul*. This is entirely accounted for by the presence of six times more heterochromatin in *P. brachiops*.[15] Large blocks of heterochromatin are visible in organisms as diverse as plants, *Drosophila*, and humans and, as we discussed in chapter 6, even simple eukaryotes with microscopic chromosomes appear to have heterochromatin at a molecular level.

Replication banding seems to be universal among vertebrates and in many other eukaryotes with chromosomes large enough to visualize.[16,17] Even in an organism such as *Drosophila*, where replication banding has not been reported on the metaphase chromosomes[10,18] (Table 7.2), the genome is clearly segregated into regions of early and late replicating DNA (chapter 6). These regions may not be sufficiently clustered to reveal themselves as metaphase bands. Even the yeast *S. cerevisiae* has early and late replicating regions of its chromosomes (chapter 5). The splitting of chromosomes into zones that replicate at different times during S phase seems a fundamental aspect of eukaryotic chromosome biology.

This sets the temporal control of replication timing at center stage in the evolution of chromosomal compartmentalization and gene regulation mechanisms.

Pachytene chromomeres are also universal among eukaryotes so far studied, thus a physical differentiation along euchromatin seems to be a basic property of eukaryotic chromosomes. This may be important for meiotic pairing[19] and/or may reflect packaging differences between euchromatin and heterochromatin or between euchromatic G-bands and R-bands in mitotically dividing cells, as in mammals.

The chromosomes of most cold-blooded vertebrates do not Q-band and this is reflected at a molecular level. The presence of base-specific fluorochrome banding corresponds well with the ability to fractionate the genomic DNA into isochores (chapter 3). The cesium chloride profiles of nonsatellite DNA obtained from a wide variety of eukaryotes vary in width and asymmetry. The degree of asymmetry, or compositional heterogeneity, is reflected in the number of sub-fractions, or isochores that these profiles can be split into. Mammals and birds, which have good fluorochrome banding, have the most asymmetric profiles and can be split into at least five isochores of differing CG content.[20] Most, but not all, reptiles, amphibia, fish have poor or absent fluorochrome banding patterns,[21] and they along with invertebrates have narrow, symmetric base-composition profiles.[20] For example, in different fish species from separate orders the clarity of the fluorochrome banding patterns was correlated with the compositional heterogeneity of their genomes.[22] The same has been found in plants[17,22], prompting the conclusion that at least some fluorochrome banding is a direct reflection in the underlying variations in the base composition of the DNA.[24] The same arguments lead to the conclusion that G-banding, that is evolutionarily more widespread, is independent of variations in the base composition.

THE MOLECULAR ECOLOGY OF CHROMOSOME BANDS

The evolutionary analysis of chromosome banding patterns, described above, suggests that the first cytogenetically detectable compartmentalization that arose in the genomes of eukaryotes was the temporal control of replication and differences in chromatin packaging (at least on meiotic chromosomes) and the segregation of some chromosomal domains into heterochromatin. G-banding (and we will assume here that this is a reflection of differences in chromatin structure on mitotic chromosomes) may have followed on from this. Fluorochrome banding seems to have appeared on the scene last of all. We have argued previously (chapters 3 and 5) that this level of chromosome differentiation may have been just a passive reporter for more fundamental processes such as replication time and/or rates of recombination and repair that had different behaviors in different chromosome band regions.

What has been the driving force (if any) for these increasing layers of differentiation between chromosome domains? It has been proposed[25] that one problem that is important to deal with as biological complexity increases is the suppression of transcriptional noise from increasing numbers of genes and from cryptic promoters within the growing amount of noncoding DNA. Quantal increases in gene number have occurred during the transitions from prokaryote to eukaryote, from single-celled to multicellular eukaryotes, from invertebrates to vertebrates as well as during vertebrate evolution.[25] Basic methods of genetic "noise-reduction" were the evolution of a nuclear envelope, to separate transcription and translation, and the packaging of the DNA into nucleosomes. This opened the door for a further level of gene control through biochemical modifications of the histones (chapter 4) and through multimeric complexes that could either interact with histones to silence comparably large regions of the genome, i.e., form heterochromatin (chapter 6) or complexes that could act as machines to displace histones to allow for transcription. The appearance of replication timing control was also apparent at this early stage. In multicellular eukaryotes these same basic mechanisms are in operation but, in addition, a growing number of proteins are recruited to perform the increasing levels of developmental control of gene silencing. For example, nucleosomal packaging may have become more versatile by the incorporation of different linker histones to this level of the chromatin fiber. Mechanisms that allowed for the differential behavior of whole chromosomes evolved, e.g., hyperactivity of the X chromosome in *Drosophila* males and somatic elimination of chromosomes in some other insects. The versatility of long-range gene silencing and activation that was necessary for the specification of complex body plans was made possible by the evolution of large multimeric protein complexes such as those involving the *Drosophila Pc*-G and *trx*-G proteins (chapter 6). Finally, in vertebrates (and perhaps in plants independently) widespread genome methylation was recruited as a dampener on illegitimate transcription.[25] One cost of this growing genome squelching may have been the need for tightly regulated genes located in quite repressive environments (such as G-bands?) to have strong regulatory elements recruited to the job of transcription. An, example of this may be the β-globin locus control region (LCR).[26]

Clearly then, there have been strong evolutionary pressures for some characteristics of chromosome domains to evolve in concert. However, it has also been suggested that no matter what selection pressure caused cytogenetically distinctive chromosome domains to come about, chromosome bands are now self-perpetuating entities.[27] Open and active chromosomal domains, as well as being more accessible for DNA and RNA polymerases, are also more accessible to other enzymes in the nucleus, e.g., those involved in DNA repair or nucleases that might be involved in retroposition or recombination. Hence the mutational input into bands of different flavor also varies. Differences in base composition and interspersed repeat content between different bands can be viewed as molecular fossils of this ordered mutation. The movement of interspersed repeats certainly appears to be karyotypically innocuous, there are estimated to have been half a million Alu insertions into the genome since human/ape divergence approximately six million years ago. This is equivalent to the movement of 150 megabase worth of DNA (the size of an average human chromosome) yet the G-band patterns of human and ape chromosomes are nearly identical.[1,27,28] One consequence of this idea that is generally not considered is that the noncoding DNA in different chromosome compartments may be evolving at different rates or in different ways. This might be clarified by long-range sequencing studies in different chromosome compartments.

A TALE OF TWO GENES

As a way of illustrating the fate of two genes that originated from a common ancestral gene by duplication, and that are located in chromosome compartments with quite different characteristics, we will compare the behavior and characteristics of the globin genes in vertebrates. The α-globin complex is located in and has the characteristics of a typical region of "open" chromatin, whereas the β-globin complex resides in a region of "closed" chromatin, in both mice and humans. Despite the fact that transcription from these two loci occurs in the same tissue-specific pattern, and that it is important that the level of protein produced from each is in balance with the other, there are major differences in the chromatin structure and behavior of the two (Table 7.3).

The human α- but not the β-globin gene has a CpG island (Fig. 3.2), whereas in the mouse neither of the two genes is marked by the presence of a CpG island.[29] The GC content of the entire human α-globin locus is in fact very high, the murine α-globin locus is still relatively GC-rich but less so.[30] Both the human and murine β-globin loci are GC-poor. The assumption is that a prototype vertebrate globin gene was characterized by a CpG island and that this has then been lost by attrition from both of the mouse genes[29] and from the human β-globin gene. A similar loss has occurred in some, but not all, of the pseudogenes[31] in the human α-globin complex (Fig. 3.2). It is not clear whether this simple story that assumes that the human α-globin gene is the most similar to an ancestral vertebrate gene is the true picture since the human α-globin gene is in fact more GC-rich and has a higher CpG:GpC ratio than either the *Xenopus* or chicken α-globin genes.[32] Consistent with the sequence characteristics of the human α- and β-globin genes, the former is located in a telomeric T-band, 16p13.3, whereas the latter is in the 11p15.4 G-band. The exact cytogenetic location of the murine genes is not known. Despite the differences in sequence characteristics of the human and mouse α-globin genes, they both appear to share similar chromatin environments and to have similar characteristics of replication timing control in relation to gene activity, and these are different from both of the β-globin loci (Table 7.3).

Table 7.3. Human and murine α- and β-globin gene complexes: a comparison

	α-globin		β-globin	
	Human	**Mouse**	**Human**	**Mouse**
Chromosome location	16p13.3	11	11p15.4	nk
Band flavor	T	nk	G	nk
CpG island density of region[69,70]	>1/100kb	nk	<1/500kb	nk
Gene density of region[68]	High	nk	Low	nk
Local %GC content[30]	60.8	55.5	40	42
Replication time when inactive[33,71-73]	Early	Early	Late	Late
Replication time when active[33,71-73]	Early	Early	Early	Early
CpG island at 5′ end functional gene[29,68]	Yes	No	No	No
DNase I sensitivity of region when inactive[33-36,68]	Yes	nk	No	nk
DNase I sensitivity of region when active[33-36,68]	Yes	nk	Yes	nk
Major 5′ hypersensitive sites[26,34-36]	1	1	4	4
Regulatory element/LCR needed for position independent expression[26,37,38]	Yes	nk	Yes	nk
Copy no. dependent, developmentally-regulated expression with LCR[26,37,38]	No	nk	Yes	nk

The characteristics of the human and mouse α- and β-globin genes and their chromatin structure when active and inactive are compared. nk; not known.

Both the cytogenetic and molecular characteristics of the human α-globin complex are suggestive of an open chromatin structure permissive for transcription even when the gene is off. By contrast, the β-globin locus looks closed when inactive but when it becomes active the region around the complex takes on the characteristics of open chromatin (DNase I sensitivity and early replication).[33] Although both α- and β-globin gene complexes have upstream regulatory elements defined by DNase I hypersensitivity (the β-globin LCR and the α-globin HS-40), these have very different properties. In both humans and mice, the α-globin regulatory element comprises one major DNase I hypersensitive site,[34-36] where as the β-globin LCR has four, which are developmentally stable and distributed over approximately 15 kb.[26] Although both regulatory elements are required to guarantee position-independent transgene expression, only transgenes carrying the entire β-globin complex are expressed in a copy number-dependent, developmentally-regulated manner.[26,35,36] Additional experiments showed that only the β-globin LCR, and not the α-globin HS-40, was responsible for setting up regional chromatin structure.[33,37,38] The α-globin HS-40 may be merely a simple enhancer aiding in directing transcription factors and the transcriptional machinery to the α-globin promoter within a permissive chromatin environment (T-band). The β-globin LCR is viewed more as an element that has a central role in determining the chromatin characteristics of an entire chromosome domain and hence, in insulating the β-globin gene from the surrounding repressive chromatin environment (G-band) in expressing cells. This role even extends to the specification of the position of replication origins within the β-globin locus.[39]

Similar locus control type regions have been found in places other than the globin loci,[40,41] and boundary-like elements have been identified in other organisms.[42] It is tempting to think of such insulating, or domain forming elements as defining topologically distinct chromatin domains, perhaps through interaction with a nuclear matrix or chromosome scaffold. There is no good evidence to support this so far. Alternatively, they may simply be elements that bind large protein complexes that act as physical barriers to the spread of repressive chromatin.[43]

THE FUTURE OF CHROMOSOME BANDING AND MOLECULAR CYTOGENETICS

A decade ago it looked as though classical cytogenetics was a field of rapidly-diminishing importance in biology and in human genetics. One technical innovation—the advent of fluorescence in situ hybridization (FISH)—has completely reversed this situation and has transformed cytogenetics from a rigid and highly-skilled discipline into the plaything of the molecular biologist. Molecular cytogenetics is in the very front line of modern biology techniques, not only in the organisms for which cytogenetics was always a key player (e.g., man) but increasingly, in the genetically-tractable organisms used in the analysis of chromatin structure (e.g., the budding and fission yeasts).

By far the widest application of chromosome banding has traditionally been in the field of human disease, and specifically cancer. Conventional chromosome banding techniques have been used to detect many structural chromosomal abnormalities, including translocations, inversions, duplications, deletions and aneuploidy. So much so that certain disease subtypes and particular prognoses have been associated with specific chromosome abnormalities, as in many leukemias.[44,45] Cytogenetics will continue to have important clinical applications in these fields.

Chromosome-specific paints[46-48] are enabling even the untrained eye to identify specific chromosomes by FISH, even those involved in complex translocations, thus reducing the need to identify chromosomes by their banding patterns. Similar paints are used in biological dosimetry to assess the effects of ionizing radiation,[49] and in evolutionary studies of comparative chromosome mapping.[50] Custom-made chromosome bar codes will make the job of precise cytogenetic location increasingly trivial and, more importantly, will allow a cytogenetic assay to be used to place new genes in relation to cloned contigs of DNA, thus helping in the integration of different types of genome maps.[51-53] These innovations on the 'wet' side of molecular cytogenetics are being accompanied by the hardware and software innovations needed to analyze complex multicolor fluorescence images.[54,55]

Another new cytogenetic technique that is proving its worth in cancer biology is comparative genomic hybridization (CGH) a powerful tool to detect nonrandom gains and losses of DNA sequences in genomic DNA.[56-58] CGH involves hybridizing equal amounts of two DNAs to normal metaphase chromosomes. One of these, the "reference DNA" is always from a normal individual, and the other is usually from a tumor from the same or an unrelated individual. After detecting the hybridization of these two DNAs using different fluorochromes, the differences in fluorescence intensity values between the two fluorochromes reflect the over- and under-representation of chromosomal subregions in the tumor genome. The advantage of this technology over conventional cytogenetics is that gains and losses of particular chromosomes and chromosome subregions can be examined in tissues, and more importantly sections of tumors, from which it has been previously difficult or impossible to obtain metaphase chromosomes. This technique is certainly illuminating the major chromosomal changes that occur during carcinogenesis and one of its main applications in the future may be in assessing the stage of a particular tumor as an aid to prognosis and the appropriate course of treatment. The disadvantage of CGH at present is that it cannot detect either sub-band size losses or gains and it cannot detect balanced translocations.

FISH in all of its guises (metaphase, interphase and extended chromatin) is going to be invaluable for the stitching-together and verification of maps of the human and mouse genomes. Will this completion of maps of the genome and their DNA sequence make cytogeneticists redundant and is it more important to know where a gene maps in relation to an established contig than to a chromosome band? The answer to this question is no in some cases if we remember that primary DNA sequence is not the be all and end all of the genetic information. The DNA sequence will not reveal anything about the types of epigenetic modifications or the higher order chromatin structure or nuclear compartment of different bits of the genome.

Knowing the normal environment a gene lies in will be important in many respects. This will help us to understand instances of position effects in human disease where genes fail to function because of their inappropriate chromosomal location,[59,60] and create position-independent transgenes. One prominent goal is to create artificial chromosome vectors for mammalian cells that might then be used in gene therapy regimes.[61] Do we want all of the genes we might wish to place into such vectors to end up in the same, controllable, environment? Or will some be expected to give high-levels of ubiquitous expression whereas the expression of others must be constrained spatially and temporally? If so, then perhaps we need to think about tailoring such vectors to have different flavors that mimic those of the different chromosome band types.

Molecular cytogenetics will also advance more and more to three-dimensional and even four-dimensional studies of in situ genome organization, including that of individual band domains. While the linear physical mapping of the human genome and other model animal and plant genomes has made great progress during the last few years, approaches to two-[62] and three-dimensional[63] and, considering the dynamics of genome organization during the cell cycle and differentiation, a four-dimensional genome analysis[64] will be indispensable to understand the in situ structural and functional organization of the human and other genomes.

At present, the field of practical light microscopy is such that we can visualize high resolution chromosome bands, but we cannot look within a chromatid, except when extra-large chromosomes are used or when chromatin structure is deliberately disrupted.[62,65] One route to improve resolution is electron microscopy but this will be at the expense of the option to use multicolor probes. The other option is to improve the resolution of light microscopes and various candidates are beginning to emerge that may fulfill this wish, including 4pi microscopy[66] and near-field microscopy.[67] We look forward to a bright and multidimensional future.

References

1. Sumner AT. Chromosome Banding. London: Unwin Hyman, 1990.
2. Herrero P, de la Torre J, Gosalvez J and Sumner AT. Heterochromatin characterisation of sex chromosomes in *Triturus marmoratus* (Urodela, salamandridae) Cytogenet. Cell Genet. 1992; 60:150-153.
3. Schmid M and Klett R. Chromosome banding in amphibia 20. DNA replication patterns in *Gastrotheca riobambae* (*Anura, hylidae*). Cytogenet. Cell Genet. 1994; 65:122-126.
4. Stock AD and Bunch TD. The evolutionary implications of chro-

mosome banding pattern homologies in the bird order Galliformes. Cytogenet. Cell Genet. 1982; 34:136-148.

5. Voloboulev Pasteur G, Ineich I and Dutrillaux B. Chromosomal evidence for a hybrid origin of the diploid parthenogenic females from the unisexual-bisexual *Lepidodactylus lugubris* complex (Reptilia, Gekkonidae). Cytogenet Cell Genet. 1993; 63:194-199.
6. Yonenaga-Yassuda Y, Vanzolini PE, Rodriguez MT and Decarvalho CM. Chromosome banding patterns in the unisexual microteiid *Gymnophthalmus underwoodi* and in 2 related sibling species. Cytogenet. Cell Genet. 1995; 70:29-34.
7. Stock AD and Mengden GA. Chromosome banding pattern conservation in birds and nonhomology of chromosome banding patterns between birds, turtles, snakes and amphibians. Chromosoma 1975; 50:69-77.
8. Wang HC and Kao KN. G-banding in plant chromosomes. Genome 1987; 30:48-51.
9. Belyayev A. Banding of G-type or R-type in prophase prometaphase in the M-chromosome of *Vicia faba* L. Caryologia 1993; 46: 301-307.
10. Holmquist GP. Evolution of chromosome bands: Molecular ecology of noncoding DNA. J. Mol. Evol. 1989; 28:469-486.
11. Greilhuber J. Why plant chromosomes do not show G-bands. Theor. Appl. Genet. 1977; 50: 121-124.
12. Viñas A, Gómez C, Martínez P and Sánchez L. Induction of G-bands on Anguilla anguilla chromosomes by the restriction endonucleases HaeIII, HinfI, and MseI. Cytogenet. Cell Genet. 1994; 65:79-81.
13. Abuín M, Martínez P and Sánchez L. Restriction endonuclease/nick translation procedure on fixed chromosomes of the Atlantic salmon fish cell line. Chromosome Res. 1995; 3:379-385.
14. Roser M. Pathways of karyological differentiation in palms (arecaceae). Plant Systemat. Evol. 1994; 189:83-122.
15. Schmid M, Steinlein C, Feichtinger W and Poot M. Chromosome banding in Amphibia. Cytogenet. Cell Genet. 1993; 62:42-48.
16. Schempp W and Schmid M. Chromosome banding in amphibia VI. BrdU-replication patterns in Anura and demonstration of XX/XY sex chromosomes in *Rana esculenta*. Chromosoma 1981; 83:711-720.
17. Kongsuwan K and Smyth D. Late-labeled-regions in relation to Q- and C-bands in chromosomes of *Lilium longiflorum* and *L. pardalinum*. Chromosoma 1980; 76:151-164.
18. Stineman M. Chromosomal replication in *Drosophila virilis* III. Organization of active origins in the highly polytene salivary gland cells. Chromosoma 1981; 82:289-307.

19. Sumner AT. Functional aspects of the longitudinal differentiation of chromosomes. Eur. J. Histochem. 1994; 38:91-109.
20. Thiery J-P, Macaya G and Bernardi G. An analysis of eukarytotic genomes by density gradient centrifugation. J. Mol. Biol. 1976; 108:219-235.
21. Schmid M and Guttenbach M. Evolutionary diversity of reverse (R) fluorescent chromosome bands in vertebrates. Chromosoma 1988; 97:101-114.
22. Medrano L, Bernardi G, Couturier J et al. Chromosome banding and genome comparmentalisation in fishes. Chromosoma 1988; 96:178-183.
23. Salinas J, Matassi G, Montero LM and Bernardi G. Compositional compartmentalization and compositional patterns in the nuclear genomes of plants. Nucleic Acids Res. 1988; 16:4269-4285.
24. Bernardi G. The vertebrate genome, isochores and evolution. Mol. Biol. Evol. 1993; 10:186-204.
25. Bird AP. Gene number, noise reduction and biological complexity. Trends Genet. 1995; 11:94-100.
26. Grosveld F, Blom van Assendelft G, Greaves DR and Kollias G. Position-independent, high level expression of the human β-globin gene in transgenic mice. Cell 1987; 51:975-985.
27. Holmquist GP. Chromosome bands, their chromatin flavors and their functional features. Am. J. Hum. Genet. 1992; 51:17-37.
28. Hwu HR, Roberts JW, Davidson EH and Britten RJ. Insertion and/or deletion of many repeated DNA sequences in human and higher ape evolution. Proc. Natl. Acad. Sci. USA 1986; 83: 3875-3879.
29. Antequera F and Bird A. Number of CpG islands and genes in human and mouse. Proc. Natl. Acad. Sci. USA 1993; 90: 11995-11999.
30. Ikemura T and Aota S-I. Global variation in G+C content along vertebrate genome DNA. Possible correlation with chromosome band structures. J. Mol. Biol. 1988; 203:1-13.
31. Fischel-Ghodsian N, Nicholls RD and Higgs DR. Unusual features of CpG-rich (HTF) islands in the human α globin complex: association with non-functional pseudogenes and presence within the 3' portion of the ζ gene. Nucl. Acids Res. 1987; 15:9215-9225.
32. Filipski J, Salinas J and Rodier F. Two distinct compositional classes of vertbrate gene-bearing DNA stretches, their structures and possible evolutionary origin. DNA 1987; 6:109-118.
33. Forrester WC, Epner E, Driscoll MC et al. A deletion of the human β-globin locus activation region causes a major alteration in chromatin structure and replication across the entire β-globin locus. Genes Dev. 1990; 4:1637-1649.

34. Higgs DR, Wood WG, Jarman AP et al. A major positive regulatory region located far upstream of the human α-globin gene locus. Genes Dev. 1990; 4:1588-1601.
35. Jarman AP, Wood WG, Sharpe JA et al. Characterisation of the major regulatory element upstream of the human α-globin gene cluster. Mol. Cell. Biol. 1991; 11:4679-4689.
36. Kielman MF, Smits R and Bernini LF. Localization and characterization of the mouse α-globin locus control region. Genomics 1994; 21:431-433.
37. Enver T, Li Q, Gale KB et al. Analysis of the developmental and transcriptional potentiation functions of 5HS2 of the murine β-globin locus control region in transgenic mice. Dev. Biol. 1994; 165:574-584.
38. Craddock CF, Vyas P, Sharpe JA et al. Contrasting effects of α- and β-globin regulatory elements on chromatin structure may be related to their different chromosomal environments. EMBO J. 1995; 14:1718-1726.
39. Aladjem MI, Groudine M, Brody LL et al. Participation of the human β-globin locus control region in initiation of DNA replication. Science 1995; 270:815-818.
40. Greaves DR, Wilson FD, Lang G and Kioussis D. Human CD2 3-flanking sequences confer high-level, T Cell-specific, position-independent gene expression in transgenic mice. Cell 1989; 56:979-986.
41. Madison L and Groudine M. Identification of a locus control region in the immunoglobulin heavy-chain locus that deregulates c-myc expression in plasmocytoma and Burkitts lymphoma cells. Genes Dev. 1994; 8:2212-2226.
42. Kellum R and Schedl P. A group of scs elements function as domain boundaries in an enhancer-blocking assay. Mol. Cell. Biol. 1992; 12:2424-2431.
43. Zhao K, Hart CM and Laemmli UK. Visualisation of chromosomal domains with boundary element-associated factor BEAF-32. Cell 1995; 81:879-889.
44. Preti HA, O'Brien S, Giralt S et al. Philadelphia-chromosome-positive adult acute lymphocytic leukemia: characteristics, treatment results, and prognosis in 41 patients. Am. J. Med. 1994; 97:60-65.
45. Walker H, Smith FJ and Betts DR. Cytogenetics in acute myeloid leukaemia. Blood Rev. 1994; 8:30-36.
46. Vooijs M, Yu L-C, Tkachuk D et al. Libraries for each human chromosome, constructed from sorter-enriched chromosomes by using linker-adapter PCR. Am. J. Hum. Genet. 1993; 52:586-597.
47. Rabbitts P, Impey H, Hepell-Parton A et al. Chromosome specific paints from a high resolution flow karyotype of the mouse. Nature

Genet. 1995; 9:369-375.
48. Guan X-Y, Zhang H, Bittner M et al. Chromosome arm painting probes. Nature Genetics 1996; 12:10-11.
49. Lucas JN, Awa AA, Straume T et al. Rapid translocation frequency analysis in humans decades after exposure to ionizing radiation. International Journal of Radiation Biology 1992; 62:53-63.
50. Wienberg J, Stanyon R. Chromosome painting in mammals as an approach to comarative genomics. Current Opinion in Genet. Develop. 1995; 5:792-797.
51. Lengauer C, Speicher MR, Popp S et al. Chromosomal bar codes produced by multicolor fluorescence in situ hybridization with multiple YAC clones and whole chromosome painting probes. Hum. Mol. Genet. 1993; 2:505-512.
52. Guan X-Y, Meltzer PS, Burgess AC and Trent JM. Coverage of chromosome 6 by chromosome microdissection: generation of 14 subregion-specific probes. Hum. Genet. 1995; 95:637-640.
53. Bray-Ward P, Menninger J, Lieman J et al. Integration of the cytogenetic, genetic and physical maps of the human genome by FISH mapping of CEPH YAC clones. Genomics 1996; 32:1-14.
54. Speicher MR, Ballard SG and Ward DC. Karyotyping human chromosomes by combinatorial multi-fluor FISH. Nature Genet. 1996; 12:368-375.
55. Schröck E, du Manoir S, Veldman T et al. Multicolor spectral karyotyping of human chromosomes. Science 1996; 273: 494-497.
56. Kallioniemi A, Kallioniemi OP, Sudar D et al. Comparative genomic hybridization for molecular cytogenetic analysis of solid tumors. Science 1992; 258:818-821.
57. du Manoir S, Speicher MR, Joos S et al. Detection of complete and partial chromosome gains and losses by comparative genomic in situ hybridization. Hum. Genet. 1993; 90:590-61.
58. Joos S, Scherthan H, Speicher MR et al. Detection of amplified DNA sequences by reverse chromosome painting using genomic tumor DNA as probe. Hum Genet 1993; 90:584-589
59. Fantes JA, Redeker B, Breen M et al. Aniridia-associated cytogenetic rearrangements suggest that a position effect may cause the mutant phenotype. Hum. Molec. Genet. 1995; 4:415-422.
60. Bedell MA, Jenkins NA and Copeland NG. Good genes in bad neighborhoods. Nature Genet. 1996; 12:229-231.
61. Brown W, Heller R, Loupart M-L et al. Mammalian artificial chromosomes. Curr. Opinions Genet. Develop. 1996; 6:281-288.
62. Bickmore WA and Oghene K. Visualizing the spatial relationship between defined DNA sequences and the axial core of extracted chromosomes. Cell 1996; 84:95-104.
63. Cremer T, Kurz A, Zirbel R et al. The role of chromosome terri-

tories in the functional compartmentalization of the cell nucleus. Cold Spring Harbor Symp. Quant. Biol. 1993; 58:777-792.
64. LaSalle JM and Lalande M. Homologous association of oppositely imprinted chromosomal domains. Science 1996; 272:725-728.
65. Saitoh Y and Laemmli UK. Metaphase chromosome structure: bands arise from a differential folding path of the highly AT-rich scaffold. Cell 1994;76:609-622.
66. Hell S and Stelzer EHK. Properties of a 4Pi confocal microscope. J. Opt. Soc. Am. A 1992; 9:2159-2166.
67. Hwang J, Tamm LK, Bohm R, Ramalingam TS, Betzig E and Edinin M. Nanoscale complexity of phospholipid monolayers investigated by near-field scanning optical microscopy. Science 1995; 270:610-614.
68. Harris PC and Higgs DR. The structure of the terminal region of the short arm of chromosome 16. In: Davies KE and Tilghman SM, eds. Genome Analysis. Vol. 6. Cold Spring Harbor Laboratory Press, 1994:107-134.
69. Higgins MJ, Smilinich MJ, Sait S et al. An ordered NotI fragment map of human chromosome band 11p15. Genomics 1994; 23:211-222.
70. Craig JM and Bickmore WA. The distribution of CpG islands in mammalian chromosomes. Nature Genet. 1994; 7:376-382.
71. Goldman MA., Holmquist GP, Gray MC, Caston LA and Nag A. Replication timing of genes and middle repetitive sequences. Science 1984; 224:686-692.
72. Dhar V, Mager D, Iqbal A and Schildktaut CL. The coordinate replication of the human β-globin gene domain reflects its transcriptional activity and nuclease hypersensitivity. Mol. Cell. Biol. 1988; 8:4958-4965.
73. Epner E, Rifkind RA and Marks PA. Replication of α-globin and β-globin DNA sequences occurs during early S phase in murine erythroleukemia cells. Proc. Natl. Acad. Sci. USA 1981; 78: 3058-3062.

INDEX

Items in italics denote figures (f) or tables (t).